Natural High

John P. Wiley, Jr.

Natural High

UNIVERSITY PRESS OF NEW ENGLAND

HANOVER AND LONDON

University Press of New England, Hanover, NH 03755
© 1993 by John P. Wiley, Jr.
All rights reserved
Printed in the United States of America 5 4 3 2 1
These essays were originally published as "Phenomena" columns
in *SMITHSONIAN,* February 1981 through January 1992.
CIP data appear at the end of the book

Contents

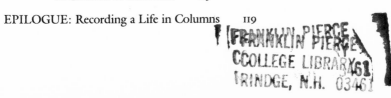

INTRODUCTION:
The Wild Is With Us Always

 We tend to think of nature as being "out there" somewhere, in the parks and preserves we have set aside so we can spend an hour or two when we get. the urge. We "do" nature like a museum: It is there when we want it. We think of ourselves as apart from the natural world, and indeed most of the time we are, living in elaborate shelters, moving from home to work to shops inside the security of our cars, working in climate-controlled buildings in which we cannot even open the windows. But nature is not "out there"—it is everywhere. Grass growing up in cracks in the sidewalk. Weeds that spring up in the smallest, meanest patch of bare dirt and the insects that seem to spontaneously generate with them. In the seams of a black granite fountain, deep green blades of grass tremble in the cascading water. Between the cut stone blocks of a seawall that keeps a river out of the city streets, woody plants take hold, growing to four and five feet before park workers cut them back one more time. Some cities are blessed with peregrine falcons that nest on the ledges of skyscrapers; most are home to nighthawks that appear in the sky as the light leaves it, catching insects we never see.

If we look, we find that nature is infinitely forgiving, that despite our best efforts to eradicate it, the birds and butterflies and beetles are not only still out there, but are pushing into our space at every chance they get. Put a tub plant on a high-rise balcony and a bird will come. Plant some reeds in a reflecting pool and ducks will nest. Invest in a specimen tree for the backyard and a beaver will fell it. Nature will survive if given even half a chance. As we live through the greatest extinction of species in the history of the planet, this one of our doing, we see reminders everywhere that we have not killed nature yet, despite the title of one recent book. Ask anyone who has put out a bird feeder. There is still time for us, a species in its adolescence, to figure out how to coexist with much older ones.

Most of us spend very little time out under the open sky, and when we are outside we seldom look up. We do not notice the dark-bottomed cumulonimbus that bring the rain, much less the high, high ice crystals of cirrus that seem to mark the stratosphere itself, as far removed from our lives as the thin clouds of Mars. Who among us can look at the sky and predict the weather? Who even remembers what phase the moon is in? How long has it been since we have seen the quick flash of a meteor? Shuttling between city and suburb, I ignore for too long the night sky because I live where there is usually not much to see. I fail to look up after a cold front has passed through, when the air is clear and the stars and planets are closer, almost over-head in a dark backyard. The evening star burns bright and I have forgotten whether it is Venus; the morning star shines in the predawn stillness, unseen.

I believe that those of us who work indoors would be happier and healthier if we spent more time outdoors, feeling the weather instead of merely seeing it. We go to extraordinary lengths to keep the weather away from us, bundling up in the cold, holding an umbrella over our raincoats as though our faces would dissolve like those in horror films if any of the rain actually touched us. Simply walking rather than driving means we will see, hear, smell, feel more. We are more likely to come alive to the moment, to experience the now instead of straining impatiently to get to the next place. We can learn to see.

Whatever their nominal subjects, all of these essays are really about learning to see the natural world around us: what is visible but un-noticed; what is there but invisible; what once was there; and what will or could be there in the future. They are reflections of 25 years as a science and natural history editor and 50 years of walking beaches and woods and looking up at the sky. At the same time they are reports of attempts to regain a childlike wonder, to laugh out loud at the marvel of it all.

In our time, learning to see means more than learning to slow down and look. That is necessary but not sufficient. Scientists have been looking, too, and they can tell of the very real we cannot see, of the emptiness of atoms and the patterns of the galaxies, as much a part of our world as the rock we sit upon. To see we need to read first, and then look and try to put it all together. Mystics tell of special moments when they feel one with the universe. All of us can use our brains, composed of matter formed long ago in the explosions of distant stars, to feel our planet wheeling around the sun, the sun around the galaxy, the galaxy on its own mysterious journey through space and time. We can lie on a chaise lounge outside on a summer night and pilot Spaceship Earth through the cosmos, or just stick out our tongues to catch a molecule of meteor dust.

To me the phrase "natural history" means all of the natural world: Rocks and soils, woodland brooks and the oceanic abyss; microbes and mastodons . . . everything. Now Webster's Third defines it as a "former branch of knowledge." Still acceptable, apparently, although rarely used, is the term "natural philosopher," meaning one who studies nature in general. We need a new phrase for the person who is interested in—and knows something about—everything. All the museums, the books and field guides, the television specials, the local clubs and national societies, popular magazines and learned journals are there for the taking. They, too, teach us to see, to know—which after all is the root meaning of the word "science." Once an educated man read the classics in the original Latin and Greek, knew history, understood in a general way all of science. How can a person be considered educated today who does not know a little something of quantum mechanics, subatomic physics, evolution, carbon cycles, plant communities, the social organization of animals?

These essays are attempts at self-education, whether walking along a river or reading, interviewing, and trying to understand. Together they become a personal ecology, a worldview in which everything is connected to everything else. Personal ecology implies understanding, and in the popular—even if misunderstood—sense of the word ("Let's recycle to help the ecology"), it implies a growing desire to live more lightly upon this Earth of ours. We begin to make connections between a Sunday walk in the marsh and what we do when we are far away. We realize that every time we walk rather than drive, or even walk up the stairs rather than take the elevator, we are easing our impact. It is the delightful situation known as "win–win." We decrease the damage we do and make ourselves stronger and healthier in the bargain (sometimes even save a little money as well).

Developing a conscious awareness of the connections between everything—us included—is the heart of personal ecology. When we fill a room with houseplants or vote for a bond issue to buy more parkland, we affect the planet. When we try to cut down on the water we use, or join the Isaak Walton League's Save Our Streams program and actually get our feet wet, we affect the planet. When we turn out a light, we cut back on electric generating demand and make the night sky infinitesimally more visible.

The essays that follow are about efforts to make these connections. Nowadays I walk a lot because heart attacks have made me nervous. But walking does more than push a little blood through my broken-down arteries. Walking makes me feel more alive, more aware of the world around me. It awakens my senses, improves my outlook on life. It helps me see.

Such a personal ecology does not require years of study, athletic training, or monastic asceticism. Learning to see, to feel more alive are not goals to be slogged toward. Rather they are a stance, an approach to the world. As soon as we try, we succeed. Then we get better. And better. The phrase may never come to mind, and if it did we probably would not mention it to anyone, but we have become natural philosophers. The funny thing is, it's fun.

PART 1
Cosmic Connections

Learning to see is partly remembering to look, and partly bringing what we know to what we are looking at. The last place we ground dwellers remember to look is up, whether at the daytime sky or the stars at night. We forget that we live not just in a town or city but in a galaxy. We read about advances in scientific understanding, but do not relate them to our workaday world. And yet we are part of the cosmos just as much as we are a part of nature.

The most spectacular thing I ever saw in the sky was a sun-grazing comet, Ikeya-Seki, in 1965. In the predawn sky it shone like a searchlight, rising straight up from the eastern horizon, a searchlight calling attention not to a new car dealership but to the origins of the solar system. It was a reminder of a time when 10-mile snowballs rained down on a new Earth.

The temptation is to wait for something equally spectacular, just as cargo cult South Pacific islanders waited for the ships to return. The trick is to look up all the time, to see the wonder that is always there. Take the binoculars you use for birding and look at where the sun is rising on the moon. On a dark night, use them to sweep slowly along the Milky Way, until you feel you are falling into an infinity of suns.

While you are looking, think of what you have read of nuclear fusion in the cores of stars, of winds of charged particles blowing off their surfaces into deep space (and into our auroras), of how astrophysicists and subatomic physicists are meeting where the largest things we know meet the smallest, the first seconds after the Big Bang. Pick up a pebble and see the atoms, the protons, neutrons, and electrons, then meditate on quantum mechanics until those particles become waves of probability, not real until they collapse on the retinas of your eyes like combers on a beach. We can all begin to feel the insights of what scientists have found; by thinking about them as we look at the world around us, we become natural philosophers.

In the same way, we can learn to follow the ecological slogan, "Think globally, act locally." The future depends on the actions of each and every one of us, just as future weather depends on the actions of each individual molecule in the atmosphere. Still another search for extraterrestrial intelligence began in 1992, on the assumption that any civilization we find is older and wiser than we are. But we cannot wait for gods to emerge from their spacecraft. It may turn out that in the entire galaxy, we are the best and the brightest. It is going to be up to us to save the only planet we have. We must become geophysiologists, concerned caregivers for an organism 25,000 miles around the waist. It starts with trying to see the world in a new way. As in the following.

Just the Right Size

 After a long day on a flat, solid and stationary world, I like to stand outside on a spring evening, watch the stars come out, and remember where we are. All day the sun beats back our gaze; it is so bright we do not look up, but only around at each other and down at the ground. Five miles is considered good visibility. Then the sun sets, and when that first star appears, we are seeing hundreds of trillions of miles.

One by one, points of light wink on. I have been watching it happen since childhood, yet something in my mind still starts when I see a star that just a moment ago was not there. The sky is still light; the first five or six come slowly, almost teasingly; then stars turn on faster and faster, gathering like timid animals that have decided they are safe. Darkness falls, and hundreds of lights dot the celestial sphere. Once again I am standing on a planet, hurtling through space, a part of the galaxy.

The clearer the sky, the more stars shine in the soft blackness. Even in a suburb, hundreds speckle the night. Binoculars show thousands. The sky looks crowded with stars. An astronomer can look at one of those points of light and know how big the star is, how old it is, what it is made of, how fast it is rotating, and what its future is likely to be. He or she can also tell how far away it really is. Long before astrophysics became a glamor field, astronomers had found how far away the stars are. They are so far away from us, and from each other, that our minds cannot understand. The stars are not crowded. They are terribly alone.

In our part of the galaxy, the typical distance between stars is about 5 light years, the 30 trillion miles or so that light would travel in 5 years. As pointed out in *The Cambridge Encyclopedia of Astronomy,* these distances are typically 30 million times the diameter of the stars themselves. If you were to take the diameter of the average person as 1 foot and spaced people as widely as the stars are, then there would be 5,680 miles—about the distance from Chicago to Ankara, Turkey—between one person and the next. All our relatives would be distant.

My average speed going to work is about 30 miles per hour. Driving day and night, with no stops at all, it would take something like 114 million years to get to the next star. At the speed of a jetliner, I would still need more than 6 million years. Even if I could maintain the 25,000 miles an hour at which the Apollo astronauts began their lunar voyages, I would need 137,000 years. And this is just to get to the next star. To calculate the time to a star 100 light years away (still close—the Galaxy is 100,000 light years across), just multiply the above driving times by 20.

On a warm evening I look up at the stars and try to sense just how far away they really are. Betelgeuse glows red at the shoulder of Orion, a star so large that we can almost make out its disk from 620 light years away. I can see it so brightly, so clearly, and I cannot get there. Ever. It is just too far away.

And this is just the neighborhood. Stars come in groups called galaxies, and the galaxies themselves come in groups, clusters and superclusters. Distances between clusters are measured in tens of millions and even billions of light years. And there's precious little in between.

Take our local group. There's us, the Magellanic Clouds (our little satellite galaxies that decorate the southern sky), the Andromeda Galaxy just 2 million light years away, and perhaps 24 others. Our little group seems to be falling toward, and may already be a part of, the Virgo supercluster of galaxies, whose center lies 60 million light years away. Not only does the space between individual galaxies tend to be empty, but we are now finding the universe to be like a foam, in which curving sheets of galaxies surround huge empty reaches of space like so many soap bubbles. These curving sheets in turn form long structures; one is now known as the Great Wall.

Living on a very solid Earth with a bright sun in the day, a moon a trifling 3 days away, and a night sky full of stars gives us a false impression of the universe. If we look at all those stars long enough, and think about just how far apart they really are, we come to realize that the universe is essentially empty.

This empty space is not exactly a vacuum. Energy may be present, and it seems that pairs of particles appear and disappear faster than the eye can follow. But in everyday terms, it is so immensely far from one star to the next, from galaxy to galaxy, and there is so little in between, that the universe is within a whisker of being empty. And if it keeps expanding, as it seems to be doing, things can only get worse.

Thinking about all that space can be dizzying (it's a little like a first cigarette), but the feeling goes away with daylight and the return of human dimensions. No matter how much empty space is out there, things are pretty solid here on terra firma. Except . . .

Except, of course, that everything is made out of atoms, and atoms are mostly empty space themselves. Our bodies, our homes, the planet itself are composed of nothing more than little clumps of spinning protons and neutrons wrapped in clouds of even smaller electrons, held together by invisible forces we cannot sense outside the lab.

Atoms are infinitesimal to begin with (100 billion billion in a drop of water), but even if we could see them there is almost nothing to see. In his book *From Atoms to Quarks,* James Trefil points out that in an atom with about the heft of an oxygen atom (say 10 to 20 protons and neutrons in the nucleus and 5 to 10 electrons in orbit), the nucleus occupies only about one quadrillionth of the atom's volume and the much smaller electrons even less. Trefil translates that "quadrillionth" this way: If you enlarge the nucleus until it is the size of a bowling ball, then this atom would be 20 miles across with the electrons, now pea-sized, scattered around the sphere. All the rest is empty space. That's what we and everything else are made of. It used to bother me to think I was mostly water (somehow the image of a jellyfish would pulse through my mind). Then I was happy to learn that my hormones alone were worth a good $6 million. Now I'm faced with being 999 trillion quadrillionths empty space. Most unsettling.

On fourth thought, though, this is the perfect spot to be in. In fact, it's the only spot to be in. We are just the right size, balanced on a knife edge between the emptiness of atoms and the emptiness between the stars. A few orders of magnitude up or down the scale, and life would be very empty indeed.

Nowadays we pride ourselves on the Copernican revolution, our reluctant concession that the Earth is not the center of the universe. We live comfortably, if forgetfully, with the knowledge that the Earth revolves around the sun, that the sun is rushing in the general direction of Hercules, that the whole Galaxy is rotating, and that our cluster of galaxies is falling toward Virgo. But in the hierarchy of distance scales, we *are* at the center (or at least in the middle), a living dividing line between atoms and stars. I suspect (on no authority whatsoever) that the stars are not the atoms of a still larger universe and that the atoms are not the stars of a still smaller universe. Whether the alternate universes postulated in quantum mechanics have different scales is a question I leave to better minds than mine. For now, we are the center. And as a scientific sample of one, I'm quite happy that we are.

May 1982

Cosmic Rays Are Fallin' on My Head

 On a warm spring evening 3 million years ago, when the only people around were australopithecines, a brilliant ball of radiation, protons, and electrons was flung out of the tortured gas cloud around the glowing core of what once had been a star. No animal, primate or otherwise, had noted the "new star" in the sky when it burned brighter than Venus after sunset. For eons the protons fled silently through the interstellar darkness, curving along the magnetic fields of the galaxy's spiral arms. A few spiraled toward the Earth. At long last one slammed into the atmosphere high over central Maryland, driving a shower of daughter particles and photons to the ground like a cosmic shot pattern 70 yards across. Some of them hit me.

At about the time the sun was coming up on the third day of the battle of Gettysburg, a nondescript orange star 180 trillion miles away experienced a small but intense flare at its surface, shooting a pulse of electrically charged particles—including some billions of protons—up and out of its gravitational well. Just 119 years later, some began striking the Earth, including my part of Maryland 40 miles from Gettysburg.

Eight minutes ago, while I was typing the word "interstellar," four protons came together in the sun and after a complicated dance ended up as two protons and two neutrons—a helium nucleus—while throwing off "little neutral ones," two neutrinos, in the process. With no mass (we think) and no electric charge, the neutrinos zipped through the seething hell of the sun's interior as though it wasn't there, and headed straight for the blue and white of Earth. And Maryland. And my head.

A split second ago, a stray uranium atom in the concrete of my house's foundation broke up, falling down the atomic ladder to become thorium and eventually lead, and aiming a relativistic electron at a spot just behind my left ear.

In the same split second an atom of radium in my skull decayed into radon, emitting a gamma ray straight into my brain. The gamma ray stripped the electrons from several atoms in a neuron and turned one mild-mannered molecule into an electrically charged free radical that ran amok until it formed an unwholesome union with an innocent bystander.

Poetic license allows me to have all this occurring simultaneously inside my head, but the phenomena themselves are all quite real. Cosmic rays strike each of us more than once a minute. About four times a second a radium atom in our own bones disintegrates. Radioactive atoms in the dirt, rocks, and water are falling apart all around us, bathing us in an imperceptible glow of alpha, beta, and gamma rays. All of this is part of the normal background radiation. It is ionizing radiation, the kind that breaks the electrically neutral atoms in my body down into positively charged nuclei and negatively charged electrons, particles looking for action. It creates free radicals, fractions of molecules, many of which are intensely reactive, positively panting to combine with something else. It can mean havoc at the molecular level, which some scientists suspect is one of the reasons we deteriorate with age.

Radio waves beeping in from the pulsars and hissing in from quasars, the microwaves from the clouds of alcohol and formaldehyde floating between the stars are nonionizing, although it's by no means certain that their sum total effect is zero. I just like to add them in when I'm daydreaming about what's going through my head—literally—while I'm sitting around. I never quite believed the woman who worked in the same building I did in New York, who said she was receiving messages from extraterrestrials through the fillings in her teeth. But I like to try to convince myself that a minor headache at four in the afternoon is the direct result not of a sherry at lunch, but of a star that blew up tens of millions of years ago, quadrillions of miles away, sending a proton halfway across the galaxy just to find me. Talk about a cosmic connection!

(At the other end of the scale, if some of the newest theories in physics are right, it may turn out that the protons in our own bodies decay into messy things like pi mesons and gamma rays. I'm still trying to sell a cartoon of two men lunching at their club. One man is holding his chest in obvious pain. His friend asks, "Angina, Charles?" and the man replies, "No, proton decay.")

I do worry that we may have overloaded ourselves by adding so many of our own sources of ionizing radiation in the span of a single human lifetime. Now we get medical X rays and the residue from weapons testing and power-plant releases. We live in a sea of lower energy, nonionizing radiation of our own making: microwave from radar and telephone relays, high-frequency radio waves from TV transmitters. (What if television can make us crazy without our even turning the set on?) Since the Second World War the government has continuously lowered the allowable limits of exposure to almost every kind of radiation, which means if nothing else that we don't know all the answers yet. And if a person dies 5 years earlier than he or she would have without all the extra radiation, there is no way to know. (Those of us who smoke don't worry much about radiation, actually; we deal in higher probabilities.)

But I'm not here to view with alarm. It's the natural radiation I'm interested in, the crossfire of invisible waves and infinitesimal particles that unites us with the rest of the universe. Even the ones that do damage are so very faint that we never feel them. But I like dealing in the very small. I once read somewhere that reducing the weight of a car by 400 pounds would improve its fuel economy by one mile per gallon. I translated that to an extra 13.2 feet per gallon for every pound of hamburger wrappers and soda cans I rake out of the car on Saturday mornings. As a pedestrian in a crosswalk I often insist that a car make its right turn before I cross: Not making the car wait with its engine running conserves fuel, reduces air pollution, and helps out with the country's balance of payments. I'm not above thinking that the argon atom I just inhaled was once exhaled by Cicero at the Forum in Rome.

So faint emanations from the stars are a natural for me. They are small reminders of large truths. We are, after all, of the stars. The iron atoms in our blood, the nitrogen in our DNA, and virtually every element heavier than hydrogen and helium was formed not in the Big Bang but in the cores of stars, which then had the considerateness to explode, sharing the wealth with us who were to come.

Many years ago I saw, at the American Museum of Energy and Science at Oak Ridge, Tennessee, a spark chamber that lit up and made a sharp cracking sound whenever a cosmic ray passed through it. In this age of miniaturization, I'd like one I could wear on my wrist. A gentle chime every minute or so would be a true message from outer space, a tie line telling me of the deaths of stars. It would be a call from home.

April 1982

Time Through a Different Lens

This old world is slowing down, and so am I. The Earth's decline is so well measured that the U.S. Naval Observatory not many years ago added a full second to a particular day, to keep our clocks in time with the planet. What slows the Earth is mostly tidal friction: all that salt water sloshing around twice a day. The day has become 2 hours longer in just the last 150 million years. For reasons not obvious to me, the rotational momentum lost by the Earth is transferred to the moon. It cannot spin any faster because one side is locked toward us, so it moves farther away and sweeps out in a larger orbit. We are getting longer days, but eventually we are going to lose total solar eclipses. The moon will appear too small to cover the sun.

I can use the extra second. In fact, I could use a lot more. What I really want is an extra month or so, right now, not another hour a day 75 million years from now. But I don't suppose that pressing a few dollars into the right hands would do any good.

My own sense of time seems to be splitting, abandoning the center for the extremes. On the macro level, time is going by so fast that the world is a blur. Seasons go by as quickly as days used to. I glance at the paper on a rainy April morning and look up in time to see the sun set after a hot June day. The car is going too fast, but there is no driver to dissuade. On the micro level, the opposite occurs. I stride down the sidewalk at a brisk city pace, and feel it takes hours and hours to get to the corner; I've fallen into some relativistic world in which time is dilating. It all probably has to do with aging, with neurons dying like shad at the end of their run, but it gives a plodding mind an excuse to shift into high.

When I go birding in a local marsh on weekends, I try to see it on as many different time scales as I can, like looking at it through a zoom lens that I can continually adjust to change the distance scales. Some things happen in "normal" time: At dusk a muskrat swims out of a marsh channel into the chop of the open river, moving with the no-nonsense economy of a workboat going out to sea as the pleasure boats come splashing in for the night. Or an osprey finds its angle and dives, and the moment is gone.

It is other velocities of time that I try to feel, the ones that require more than an hour's stroll—or a human lifetime. The tides rise and fall invisibly, so gently it is difficult to think they slow the Earth. I stare at a slab of broken concrete in the water and strain to see the water rise. A pebble in the conglomerate is my marker. But it always happens when my eye blinks: One second the pebble is dry, the next it glistens wet. For me the cycle from high tide to high tide takes about 13 hours. For this great hump of mud in front of me at the water's edge, as full of life and as shiny as the swollen abdomen of a queen termite, the tides may be the systole and diastole of its circulation, and 72 high tides make one of its minutes as 72 heartbeats do one of mine.

Just as I cannot see the tides move, even when they trap me out at the end of the point, I cannot see the plants grow. Green reeds shoot up through the dead stalks of last year; wildflowers wink into existence along the trails. Time-lapse photography of flowers opening is famil-iar, revealing motion we could not otherwise see. But I want to see more.

Images of tropical vines tantalize me. Thomas S. Ray, Jr., described vines germinating on the forest floor, snaking along the ground until they encounter a tree, and then climbing the tree until they reach the light they need to flower. If the tree encountered is not tall enough to reach the light, the vines turn right around, go back to the ground, work their way across until they encounter another tree, then try again. No part of the vine actually moves: The front end keeps grow-ing while the back end keeps dying off. Any given part of the vine is stationary, one definition of a plant, and yet the organism manages to move. If we could accelerate time, we would see the vines moving through the forest like so many green, leafy snakes. Lacking that power, we can try to alter our perception of time. If it happens accidentally, as has been the case with me recently, perhaps we can learn to induce it.

I want to achieve the time sense of a sycamore that might live 600 years along the shore, and feel the seasons come and go like so many days and nights. Years might be as days, alternating light and dark, warmth and cold; perhaps only the fierce storms that come once in a generation might stand out enough to mark the passage of time. Then I might see the vegetation move, rising on the incoming tide of spring and falling back on the ebb in autumn. I might see the marsh itself move as mud flats become solid ground and river sediment makes new marsh.

Somewhere beneath the mud are the rocks, and they have their own time sense, one in which a marsh or a civilization lasts only a moment. Like molecules of water that evaporate from the sea, float across the sky only to fall as rain, and wash back to the sea again, the rocks have pushed up out of the planet's molten interior through the hole in the bottom of the sea, and are bumping along on top until they plunge back into the warmth from whence they came. Crustal dynamics give us earthquakes and recycle the planet's skin.

Plate tectonics is not the first thing that comes to mind standing in a marsh, and yet my marsh is sliding west, away from the Mid-Atlantic Ridge. On the theory that what comes up must someday go down, I had once assumed that the marsh—along with the rest of North America—would one day be subducted under some other plate, and the slate would not only be wiped clean but melted down. Not so, it seems. Continental plates are lighter than oceanic plates, and tend to ride up over the latter. Neither the marsh nor the continent is going to be recycled. We have to live with what we have.

At worst it is pretentious of me to work at seeing the marsh in accelerated time. It can be comical: The reeds spring up and die back in a rhythm that resembles nothing so much as those early black-and-white movies in which people seem to run in quantum spurts. It can be dead wrong, as was my vision of North America going under. And yet there must be some underlying truth to be known, however dimly, if we can learn to tune our perception of time as we do our perception of distance.

At a junior high school concert some years ago, the chorus director brought out a shy boy, younger than his schoolmates, to play the piano. She had accidentally overheard him play once, and wanted us to hear him. He played a classical piece and he played it fast, incredibly fast, faster than anyone I ever heard. He played so fast that great splashes of notes became single tones in their own right, tones that combined to produce a simple, clear melody, a song by itself floating over the torrent of notes bursting from the piano. It was a magical song, one that must have existed in the composer's head, the unwritten and perhaps never-before-heard soul of the composition.

In the marsh, the birds and turtles and dragonflies and I go about our business in normal time. But the tide covering a pebble can split my time sense, and the marsh and life itself move faster and faster. And I sense that a song is being played, a song just out of my hearing, the song of life.

July 1983

The Universe as a Hologram

 When I was a backyard astronomer, I used to lie in a chaise longue outside at night and try to feel the motions of the universe. I would imagine I felt the Earth rotating to the east at a modest 750 miles an hour or so. Then I would figure out where the sun was under me, and try to sense the Earth moving at 18 miles a second along its orbit. My stomach would drop as the solar system plummeted toward Hercules. I would find the Milky Way, the equatorial belt of our own galaxy, and join its 200-million-year rotational waltz. On a good night I would have a high time just looking at the stars.

Nowadays I look closer to home, but find myself becoming even giddier, and with good reason. Instead of looking at a star, I look at a tree. And promptly fall through the glass of our everyday world into a whirling void that makes me dizzy.

We all know in a textbook, give-it-back-on-the-exam kind of way that the tall, leafy object we perceive is really nothing more than a lot of empty space occupied in a regular way by protons, neutrons, and electrons, exactly the same particles that comprise a rock, a baby, or a star. And we're aware in a vaguer way that these elementary particles themselves are made up of still smaller and less graspable particles in turn. But lately I've been doing some reading, and it seems the situation is getting out of hand. The tree may not be there at all, as a tree, until I or you or someone perceives it as a tree. And that "tree" and I or you may be part of the same one wholeness that fills the universe.

For some centuries we have thought of nature as separate from ourselves, something we can take apart and understand like a watch, a machine. Not so, apparently. It's not just that the tree is a fog of subatomic particles too small for us to see. It's that the tree is really composed of waves of probability—potentials—that don't even become particles until they encounter a detector, such as our eye, and the waves collapse into something real.

The old saw about a tree falling in the forest when there is no one around to hear it is true. Acoustic waves there may be, but with no ears to convert them, there is no sound. The tree does not even have to fall to raise such a question. Quantum mechanics tells us that there is no tree at all until someone sees it. The tree is some combination of whatever potential is really there, our physical perception of what is there, and what our minds make of it.

The hologram is the analogy of choice, and it holds up well. Holograms are a science writer's bread and butter: high-tech (you need a laser to see them) and full of possibilities. In essence, a hologram is a photographic plate that captures not an image but the interference patterns of light waves coming from an object. Shine a laser through the plate, and the object is reconstituted in space before you in three dimensions, not the two dimensions of a photograph. Move your head, and you see a different view of the object.

What's most remarkable about a hologram is that every part of it contains all the information about the whole object, just like the DNA in every cell of our body contains the blueprint for the whole thing. Break a hologram in half, shine a laser through just that half, and the whole object is seen. Our brains seem to be holographic in storing memories. Memories are stored in every part of the brain. Destroy one part: The memory survives in others.

Now it turns out that the whole universe may be a single hologram: The information about all of it is encapsulated in every part of it. Including you and me. This idea is not just pretty theory, the sort of thing that trickles down to cocktail party talk. This idea appears to have been proved in at least five separate experiments in the last decade. In fact, at the moment, there's no way out.

The experiments have to do with creating pairs of particles, each of which then flies off in opposite directions. Each particle seems to know instantaneously what is happening to the other. If two objects are separated in our ordinary space-time, it should be impossible for information from one particle to reach the other at a speed faster than light; but it happens instantaneously. Quantum theory had predicted it. Einstein had pointed out the paradox in a thought experiment; now it is really happening in the laboratory. The short answer is that sub-atomic particles are not particles at all in our everyday sense of the world. They are things not quite real that exist in a quantum world. The long answer is light years beyond me. The point is that all the particles in the universe are, in some sense, connected.

Wholeness is the key, and my reading has led me to a man whose ideas about wholeness not only make sense out of quantum physics but have remarkably much to say about such "real-world" concerns as politics, economics, and society. He is the late British physicist David Bohm of the University of London, author of *Wholeness and the Implicate Order*.

Instead of starting with the fragments of our "real" world that we can actually see, such as my tree, or even an atom, Bohm's ideas begin with the whole universe as a constant flowing, a flux. In this river, the currents from time to time produce little vortices, little whirlpools that we can see. These are the fragments of the sensible world that we do see, trees and galaxies. He sees the universe as full, as a sea of energy in which the ripples on top are the pieces of our world, and where the Big Bang was merely the coming together of ripples to form a high wave, which then spread out in the swelling rings of an expanding visible universe.

This flow, this "holomovement," as Bohm calls it, is some kind of higher-dimensional reality, in which everything we know is enfolded like the images in a hologram. When we look at a tree, we are seeing an unfolded aspect of this higher reality.

Talk of higher dimensions has always made me feel cheated, left out by some mathematical trick. But Bohm offers a nice device for thinking about it. Imagine an aquarium with a fish swimming inside. Two television cameras are aimed at the fish through two of the sides, at right angles to each other. You are looking at two TV screens. In one the fish appears head on, in the other you see it from the side. Each of these images is a two-dimensional projection of a three-dimensional fish. The real fish is of a higher dimensionality than the two projections: The three-dimensional reality holds these two dimensional projections within it.

Just so, we can think about the particles we talk about in physics as only the projections of higher-dimensional reality. Under ordinary conditions, it works perfectly well to treat them as real in themselves. But, as we are being forced to concede by quantum mechanics, not always. In quantum mechanics we have to deal not only with the problem of separated particles acting as though they were connected. We also have to live with being forced to describe an electron sometimes as a particle and sometimes as a wave, depending on what we are doing at the moment. We even have to deal with discontinuous movement, as when electrons move instantaneously from one orbit to another without passing through the space in between.

We have an even deeper problem, Bohm explains: While classical physics works fine in our everyday world, we need both relativity and quantum mechanics to explain our universe at the levels of the very large and the very small. Both work nicely, and have become indispensable, but they contradict each other. Einstein didn't turn his back on quantum mechanics out of simple pride of authorship—he couldn't stand the contradiction. Physicists go about their work perfectly conscious of these paradoxes, and of course, they are right: We can only work with what we can "see." But ignoring the fundamental problem doesn't make it go away.

Bohm's underlying flowing wholeness can resolve the contradictions. If we start with the premise that when we "look" at an electron, we are seeing only a limited aspect of a higher-dimensional reality, we can expect to see it sometimes head on and sometimes sideways. We can expect to see a vortex in a stream disappear here and reappear there. If we think of our theories not as reality in themselves but both as a way of explaining what we see and as a clue or an avenue to a deeper reality, we will then keep progressing to better explanations of our sensible world and at the same time keep approaching the deeper reality.

Bohm's book is not only about physics. He wants us to become aware that our view of our own world has become fragmented, especially in the sciences but also throughout our everyday lives. We divide our universe up into stars and atoms, and we divide ourselves from nature, and ecosystems from each other. We divide ourselves into races and nations, into political and economic fragments, and deny any underlying wholeness. He says, "It is not an accident that our fragmentary form of thought is leading to such a widespread range of crises, social, political, economic, ecological, psychological, etc., in the individual and in society as a whole. Such a mode of thought implies unending development of chaotic and meaningless conflict, in which the energies of all tend to be lost by movements that are antagonistic or else at cross-purposes."

February 1981

We Are the Extraterrestrials

Something deep within many of us wants not to be alone. We not only like movies about extraterrestrials, we secretly hope they will land in our neighborhood, arrive at our house. We are embarrassed by the more lurid UFO tales precisely because we want so much for the next one to be true.

More than once, standing in a field at night looking up at the stars, I have found myself thinking, "Now would be a good time." I watch for a light that grows brighter, comes closer, swoops in for a landing all unseen by neighbors or military radar. Too many bad movies too long ago, perhaps. For a few minutes one night I thought I had my UFO. Driving along a busy interstate highway, I saw a light in the sky, moving slowly and exhibiting a pattern of brightening and dimming. I pulled over, stopped in a shower of gravel on the shoulder, and leaped out of the car. The light drew closer and closer, until I could see that it was a small plane, advertising the county fair with an array of lights on the undersides of the wings.

Many people, I suspect, want the aliens to arrive for the sheer excitement and for the promise of escape from the bonds of daily life. (People have welcomed the outbreak of war for the same kinds of reasons.) Even before we know the aliens' intentions, we are sure that tomorrow will not be like today and it is easy to feel that any change is good. The problems that weigh us down are suddenly insignificant, and we are caught up in a larger issue, the largest the world has ever faced.

We have a history of yearning for alien life. The popular response today to movies about lovable, or at least benign, aliens seems little different from the excitement of newspaper readers in 1835 over a phony report that Sir John F. W. Herschel had "discovered" batlike people living on the moon. The hoax could not have lasted as long as it did unless people wanted to believe it. (Herschel himself was in Capetown at the time and knew nothing about it.) Percival Lowell, seeing the canals he wanted to be there, was not alone in adamantly wishing for life on Mars.

And the yearning may be for more than simple excitement. As individuals and as nations we are not generally doing as well as might be hoped, and yet it is not clear how to do better. We live with a sense that we should be doing better, a kind of "divine discontent" once used as an argument for the existence of God. Extraterrestrials arriving in starships would be so far ahead of us technologically that we could hope, even assume, that they were ahead of us socially as well. At the very least they would have avoided destroying themselves and been able to cooperate well enough to travel the stars. Presumably they would have something to teach us about living together as well as about rocket engines. For us they would be a deus ex machina.

On our better days, when we are pretty sure we can muddle through without any help from anyone, we can still prefer to believe that there is life elsewhere in the universe. Otherwise the waste is overwhelming. One hundred billion stars in our galaxy, 100 billion galaxies: A universe billions of light years across and just one inhabited planet? It just doesn't compute. (To fully appreciate this argument, step outside some dark, moonless night and slowly sweep along the Milky Way with a pair of binoculars. And that's just a small part of one galaxy.)

Some of the intellectual arguments, as opposed to emotional longings, for life elsewhere depended on nothing more than numbers. Harlow Shapley, the late Harvard astronomer who first nailed down the dimensions of our own galaxy, was forceful in his book *The View from a Distant Star:* " . . . let us say that only one star in a hundred is a single star, and of them only one in a hundred has a system of planets, and of them only one in a hundred has an Earth-like planet, and only one in a hundred Earth-like planets is in that interval of distance from the star that we call the liquid-water belt (neither too cold nor too hot), and of them only one in a hundred has a chemistry of air, water, and land something like ours—suppose all those five chances were approximately true; then only one star in ten billion would have a planet suitable for biological experimentation. But there are so many stars! We would still have, after all that elimination, ten billion planets suitable for organic life something like that on Earth."

Shapley's worst-case scenario would provide only 10 such planets across the hundred thousand light years of our own galaxy. One gets the feeling he thought the real chances were a good deal higher. In fact, in a paper published just a year before that book, he proposed that life could arise on planets not connected to any star but floating in space, self-heating in the way Jupiter is.

There are lots of sophisticated arguments for extraterrestrial life, beginning for me with Walter Sullivan's book, *We Are Not Alone,* and the collaboration of I. S. Shklovskii and Carl Sagan, *Intelligent Life in the Universe.* Not everyone, of course, is convinced, just as many people would just as soon that a spaceship not land in their backyard. And lots of scientists think they have the answer to Enrico Fermi's demand: Where are they?

The biological argument was summarized by Frank J. Tipler, an associate professor of mathematical physics at Tulane University. According to Tipler, the best contemporary thinkers on evolution—Francisco Ayala, Theodosius Dobzhansky, Ernst Mayr, and George Simpson—all maintain that "the evolution of an intelligent species from simple one-celled organisms is so improbable that we are likely to be the only intelligent species ever to exist."

In the same article Tipler contends that if intelligent life had arisen elsewhere, it would have spread through the galaxy and would already have arrived here, at least in the form of self-reproducing computer probes. None has arrived, again raising the question: Where is everybody?

So perhaps we are really alone after all. No one is coming to bail us out, no one is coming for a visit, no light will ever swoop down to my field in the night. No god will pop out of a strangely lit machine. Our future is entirely up to us. If we make it, we will have done it by ourselves, and will be able to take appropriate pride.

But there is another possibility. Suppose we are not the only, but the first? Suppose that in galaxies across all of space and time, as stars formed and synthesized heavy elements and exploded and died, as planets accreted out of the debris, as organic molecules became more complex in an evolution toward life, we were the first? Not in the whole universe, perhaps; modern cosmology weighs powerfully against any argument for the Earth's uniqueness. But suppose that, in the random order of things, life did arise first on Earth in this galaxy, or in this sector of the galaxy. And suppose there is life on lots of other planets around us, but it is a million or a hundred million years behind us.

Then we can imagine that someday in the distant future the life forms on those planets may find themselves discontented and yearning for something to happen, looking at the night sky and hoping for company. Starships will appear in their skies, and the aliens will land. They will have something to offer about rocket engines, and about how to make a go of life. They will be us.

January 1983

Leaving Earth to Save It

A dark forest used to stand silent just inside the National Museum of Natural History, a growth of hemlocks 4 to 5 feet in diameter. Through the trees you could see a river, a wooded island on the right, and acres of wild rice to the left. No bridges crossed the river. It was the spot where Rock Creek empties into the Potomac River in Washington, and you were seeing it as the Indians saw it before Europeans arrived. Malls and memorials stand now where once the wild rice grew; a seawall keeps the river in its place.

When I walk the seawall, I try to see the river as it was. I do the same on Chesapeake Bay or along the Hudson River, remembering accounts of the extraordinarily abundant fish and wildlife found by the early explorers. And I wonder how well they might recover if we went away for a few centuries. Would hemlocks grow again in what is now a boathouse parking lot at Rock Creek?

To find out what would happen, one could consult the literature on plant succession, study abandoned highways, visit ruins in Mexico, travel to lost cities in Asia. Or one could make evacuating the entire Earth the premise for a science fiction story, and let the special-effects people fill in the details when the story is snapped up for a major motion picture. New York City 10,000 years after the last human left would be a new challenge for the model makers.

The scenario goes something like this: The time comes when, despite our best efforts, the only way left to save Earth is to leave it. Everybody. For a long time. So many species have been lost, so many ecosystems impoverished, that the whole biological life support system is close to collapse. The natural waste-removal systems, the recyclers, the air filters and water holders are being overwhelmed.

In this fantasy future, field biologists are the new elite. They are paid more than Congressmen, although less than basketball players. Biologists have been multiplying as fast as species have been disappearing. In 1980 one expert had told Congress that there were only 1,500 people in the world competent to identify tropical organisms; when brown leafhoppers destroyed several billion dollars worth of rice in Southeast Asia in the late 1970s, only a dozen people in the world could distinguish with certainty the 20,000 species that make up that insect group.

Now, decades later, armies of biologists carry on with a wartime intensity, desperate to learn more about how the natural systems work before they disappear. They need to know not only what should be saved first on Earth, but what should be added to the recycling systems on the space colonies and asteroid mines overhead, which, like most home aquariums, do not quite work as closed systems. By now humans live as far away as the moons of Jupiter, but all the secrets of life remain on Earth.

At some point it becomes clear that the race is being lost. True, human numbers are dropping. Heavy industry has moved into space. But so much tropical forest has already been cut, so many watersheds destroyed, so much topsoil washed away that the biological decline has unstoppable momentum. It is too late for management, no matter how wise. Thus a decision once made by bands and tribes, to pack up and move, is now made by the population of a planet. The actual mechanics are a little fuzzy in my fantasy, except that more and more people would move into space to build more and more colonies for more and more people to move into space.

With a little suspension of disbelief one can see the story unfold on the Earth they left behind. Grass appears in the boathouse parking lot. A small section of the seawall collapses and, with no one around to fix it, the river moves in like a silent bulldozer. The noblest experiment of all has begun. The story moves in fast-forward, time-lapse photography; we watch nature reclaim itself through the eyes of appropriate creatures. Early on we see city streets through the eyes of a rat; we follow the feral dogs and cats that roam the suburbs along with the

29

raccoons and skunks. A century later a coyote hunts in the streets of Manhattan; five centuries after that a panther uses the pinnacle of a rubble pile to search for prey. A pigeon's eye view of a city changes to a falcon's.

Humans will not be able to leave the planet completely alone, of course, any more than an editor can pass on a manuscript without making a mark. The luckiest biologists of their generations will be landed at monitoring stations. Remote sensors will be maintained, camera lenses cleaned. A little crisis intervention might be allowed at first: aerial tankers putting out a fire about to destroy the last known stand of some special plant. But as far as humanly possible, the biologists would keep their hands off, acting only as passive receptors. Nature would be the protagonist and the star. A living Earth regenerating from the ruins. The miracle that struggles to happen in every vacant lot happening everywhere. All with appropriate inspirational music, of course.

Conflict for a story line should be easy. At the start the conflict would be over whether to leave; I could stoop to a line something like: "Gazing out into the dark—30 stories above the East River— after a day of hearing out the ecologists' deputations, the Secretary-General became biologically literate." Later the conflict over whether it was time to return would grow stronger. The latter argument would not sound entirely unfamiliar to anyone who follows today's debates over multiple use versus preserving wilderness areas unsullied. A writer might suffer the temptation to inject ideological harangues into the dialogue.

In the meantime, the miracle would be fact. The biological crisis would be past. Nothing would have returned from extinction, but the several million species left would be plenty to keep the bio in biosphere. An optimist would end the movie with humans returning to live in gentle coexistence. A pessimist might have us come back to ravish the Earth all over again. Something in between seems more reasonable.

I'm completely over my head in every part of this fantasy, of course. If we flew 100,000 people off this Earth every day, it would take 125 years to move the current population. I certainly don't know if hemlocks could or would ever grow again in that boathouse parking lot.

But I like the uniting of what are now inimical factions: the high technologists who believe the future of our species lies in space, and the environmentalists who fear we would foul the rest of the solar system just as we have fouled the planet. Environmentalists might be a little humbler if the spacers save the world. I also like the idea of a species, grown out of its infancy, given a second chance at husbanding a remarkable place to live. Possibly even the only place to live.

But the strongest appeal is really the vision I started with: Fish leaping in clear water, forests growing to the water's edge, swamps and marshes pulsating with life. The way things were here just five or six lifetimes ago.

To some people today, an environmentalist is a monomaniac. Worse, still, is a preservationist: An elitist unconcerned with people. Perhaps I am most unspeakable of all: a preservationist who not only wants to keep what we still have but would like to bring back what we once had. I don't feel antipeople at all. We people need the life support of the biosphere. The whole system is slowly failing, but still has the power to regenerate without any help from us. All we have to do is stand back. Now, if only that were somehow possible.

September 1982

PART 2

Perturbing the Natural World

Dioxin falls with the snow on Antarctica. Dust particles from tea processing in Kenya provide the nuclei on which water droplets condense, changing rainfall patterns. Cities stand where grizzlies once fed on dead whales on the shores of San Francisco Bay. Nature may not be dead, but it is hard to find.

We can see (second growth) forests being turned into farms, and farms into housing tracts, office parks, and shopping centers. We see silt in rivers the color of coffee with cream, and trash lining the banks. We cannot see gases coming out of tailpipes or used motor oil seeping through the ground. And we cannot see, and cannot remember, what was here before today. This last is not just a lament for the extirpations of history, for the wolves in fur that no longer prowl the Potomac. It is a question of racial memory: When we are out birding in the local park, we do not miss the migratory species that were there less than half a human lifetime ago because we have never seen them there.

Sometimes we are like children, not understanding the consequences of our actions. We leave lights burning, and lose a little more of the night sky. We fertilize and spray our gardens, and poison an estuary a hundred miles away. We casually use the products of modern technology, releasing into the atmosphere chemical compounds that turn precipitation acid or thin the ozone layer overhead that protects all life.

Oftentimes, of course, we do realize the consequences, and people of good will can bitterly disagree, as in the fight over whether a mountaintop should remain a biological refugium or become a world capital of astronomical research. Clear, dramatic choices are the exception, however. It is the little decisions each of us makes every day that will decide the fate of the Earth. Only a revolution in our thinking can change the course of what we are doing to our planet. But it will be an invisible revolution, of the kind described by Marilyn Ferguson years ago in *The Aquarian Conspiracy*: small steps taken by millions of people, each step leading to another. People will move ever so slowly away from the three C's that James Lovelock, the exponent of Gaia, says are killing organism Earth: cattle, cars, and chainsaws.

Just as it is hubris to think we can "conquer" nature, however, it is equally presumptious to believe that only we can save it. The planet survived bombardment by asteroid and the introduction of oxygen into the atmosphere at a time when oxygen was poisonous to the anaerobes that were then the only living things. Islands survive hurricanes; continents survive ice ages. On a smaller scale, biologists have learned that seemingly homogeneous habitats are actually mosaics. Anthills "disturb" prairies, bringing up subsurface minerals in the process. Treefalls "disturb" the woodland canopy, letting in a patch of sunlight and setting off an explosion of growth on the ground. Some ecosystems must have fire; others depend on floods. Disturbances are natural, even necessary; we need to understand while we do what we can to undo what we have done.

The Heavens Lost in Our Glare

 Looking at the night sky is like going birding: Every year there is less to see. Unlike warblers and vireos, the stars are not actually disappearing; rather, we are slowly being blinded. The air we look through is a thin soup of water vapor and aeroplankton, dust and hydrocarbons. But it is not air pollution that blocks our vision; instead, the light coming to us from near and far corners of our galaxy is being overwhelmed by the light we generate ourselves. Our own nighttime illumination is blinding us to the light of the cosmos. Textbooks still say 6,000 stars are visible to the unaided human eye. The Milky Way is portrayed on sky maps as a prominent feature, as though the reality were a fine spray of white on black. For most of us, that's history.

"Deep-space" astronomers, those patient souls striving to see to the edge of the Universe and back to the beginning of time, are hurt the most. They are trying to see objects so dim, so far away, that at times their instruments are counting individual photons that have been crossing the intergalactic darkness for most of the lifetime of the Universe. It so happens that we just now are discovering things about the structure of the Universe as revolutionary as our realization early in this century that our home galaxy is just one of innumerable island universes. We found great voids in what had seemed the random, even distribution of galaxies. Then we found that galaxies cluster on the perimeters of these voids, forming a "soap-bubble" universe. Now we are seeing a "great wall" of galaxies, the largest structure ever seen. The discoveries are coming thick and fast, and all the while our great telescopes are in effect becoming smaller and smaller.

David L. Crawford is an astronomer at the Kitt Peak National

Observatory near Tucson, Arizona. Also a leader in the campaign to protect observatories from light pollution, he is both chairman of an International Astronomical Union (IAU) colloquium on the subject and founder and driving force of the new International Dark-Sky Association. He describes the latter as a "nighttime Sierra Club—but not radical." Crawford estimates that the 200-inch Hale telescope on Mount Palomar in Southern California has been reduced to a 139-inch instrument by the encroaching lights of San Diego County. Even the telescopes in the high desert of Chile are losing their power, Crawford says, to the lights of growing La Serena, 50 miles away.

For amateur astronomers, the problem is just as real. Consider the catalog of Charles Messier, called "the Comet Ferret" by King Louis XV for his discovery of a score of the ethereal interlopers from a rooftop tower in Paris two centuries ago. To save himself time wasted on comet look-alikes, he put together a list of fuzzy objects in the sky, most of which we now know to be globular star clusters, patches of glowing gas, or other galaxies. Other observers added to the list. Today amateurs like to work their way through the catalog: There is a club for those who have found all 110 Messier objects.

It is no simple task. A colleague of mine at *Smithsonian Magazine* has carried his 8-inch telescope to parks and beaches, the darkest places within easy reach, in a year-long hunt for the Messier objects. Even with equipment to aim his telescope at the precise location of a given object, he found that some were almost impossible to see. Yet his reflector has three times the light-gathering power of the instrument Messier used in the heart of a capital city.

Astronomers are not the only losers through what Crawford calls the "trashing of our night environment." In his introductory talk to the IAU colloquium, he argued:

> Mankind's view of the Universe is worth protecting even without astronomy. Such a view, from a prime dark site, is one of nature's most wonderful marvels. Few have seen such a view and not been deeply impressed. Face it, astronomy is a philosophy and an art as well as a

science. Considerations of the Universe, what it is, what it means, what's out there, are fundamental. Without the view of the Universe, these are lost. And, if lost, mankind's sensitivity to the Universe and the environment will continue to decline.

Crawford believes that, in general, things will only get worse.

The impact will be most severe, of course, on future generations. They will face the specter of not ever seeing the Universe "live," but only on a planetarium screen or on TV: In fact, that is the only way that many now see nature—on TV. There is no question in my mind that this lack of sensitivity to the environment is at the root of mankind's apparent effort to destroy that environment, taking all species, including mankind, down as well.

For the professionals, some of the news is good. Ordinances to control outdoor lighting have been enacted around the country. San Diego, whose lights threaten Palomar, recently switched from high-pressure sodium lamps, which ruin spectroscopic observations, to low-pressure sodium, which does not. The city, incidentally, will save $3 million a year, Crawford points out.

The worst problem is outdoor lighting: sports stadiums, auto dealerships, and lights aimed upward at billboards. At least 30 percent of this light misses its target and is wasted. Or, as Crawford puts it, the United States spends a billion dollars a year "lighting up the underbelly of airplanes." Translated, that sum comes to 6 million tons of coal or 23 million barrels of oil.

The solutions are as simple as using fixtures that direct all of their light downward or directly at the target, choosing sodium over mercury lamps, installing timers that shut lights off after midnight when few people will see what is being illuminated. It means saving money, stretching resources, alleviating air pollution. It even means a break for wild creatures disrupted by lights, whether they be moths pollinating night-blooming flowers, birds migrating, or sea turtles looking for a beach on which to lay their eggs. For those who care to relish the cosmos, it means reclaiming the night. As Crawford says, cutting down on light pollution is a win—win situation.

We are not all county commissioners busy buying streetlights and passing ordinances. But we do have access to light switches. And all the light escaping our homes adds to the loom in the sky. One little light does not seem like much until you spend some late-night hours in the backyard, maneuvering around the tube of a trusty 6-inch reflector. Then the sudden switching on of a bathroom light nearby is like a floodlight coming on. Night vision, which took 20 minutes to acquire, is gone in an instant. And you start to think about all the lights burning in empty rooms.

Thinking about these things brought on old urges, so one night in late February I left work well after dark to see for myself. All day a 20-mile-per-hour wind from the northwest had been bringing clean air down from Canada (never mind what a southwest wind sends up there); the sunset had been cloudless. The waning moon was all the way around to Sagittarius in the morning sky. The night would be as good as it gets in the nation's capital for stargazing.

Down the Mall I went, past the joggers and the homeless wrapped in blankets. Up the hill to the Washington Monument and down the other side into a dark area that faces the Reflecting Pool and, six blocks west, the Lincoln Memorial. The air was so clear that I could follow for miles the lights of planes climbing north out of National Airport.

Above, Jupiter burned near the zenith. Orion the hunter bestrode the meridian, followed by his faithful dogs with their bright stars, Sirius and Procyon. The Gemini twins, Castor and Pollux, rode higher in the sky. And that was about it. The three stars of Orion's belt were easy to see, but those of the sword were at the limit of visibility; the fuzzy patch of nebulosity, M42 in the Messier catalog and now known to be a region where stars are forming, was quite invisible, although it is considered a naked-eye object. Aldebaran, the red eye of Taurus the bull, was easy, but I could see only three of the Seven Sisters, the Pleiades. To the northwest other stars forming the squashed "W" of Cassiopeia were just visible, but there was no sign of the Milky Way that spills over it. It took me a long time to find the North Star. My *Sky & Telescope* star map showed that the Andromeda galaxy (M31) was still well above the horizon in the northwest, but there was no sign of any of the stars that might have led me to it. In the language of ecology, this was a depauperate sky.

It is more and more the case that birders have to travel to see birds that once were common. The same holds even more so for amateur astronomers—or anyone who wants simply to see the individual suns that pull our eyes and minds deeper and deeper into space, and the broad band of light that marks the great wheel of the Milky Way, our galactic home. Nowadays (nowanights) it means either driving out of town just to look up, or waiting for those increasingly rare occasions when our travel on other business serendipitously presents us with a dark sky. Resorts in the Southwest now advertise their night views to frustrated astronomers in other parts of the country: They are selling what used to be free.

Crawford's "win–win" argument makes sense. If the flip of a switch can save us money, lessen the poisoning of the air we breathe, help one wild creature make it through the night and give us back a glimpse of what Messier saw, it would seem worth considering.

April 1990

Fighting for a Sky Island

Sometime during August [1988], the U.S. Forest Service is going to try cutting a baby in half, certain it will not be cheered as another Solomon. The regional forester in Albuquerque will decide the bitterly disputed future of a sky island in Arizona, a mountaintop where a taste of Canada has hung on since the last ice age. This is not the usual fight between real estate developers and preservationists, but one between scientists who see the mountain as a site of international astrophysical significance and a remarkable coalition of people who normally would not talk to each other but who all want the mountain saved as a unique biological site.

Leading the astronomers is a former submarine sailor who acquired a Greenpeace T-shirt while fighting for whales in Hawaii; leading the other side is an air-conditioning contractor who has put together an opposition in which animal rights activists and Earth First!-ers join hands with hunters and members of the National Rifle Association.

The issue is whether to build a world-class astronomical complex atop 10,700-foot-high Mount Graham in the Coronado National Forest east of Tucson. Popping up nearly 7,000 feet from the desert near Safford, it is the northernmost limit to the range of many plants and animals native to the Sierra Madre Occidental of old Mexico to the south and the southernmost limit of many Rocky Mountain species. It is home to more than a dozen endemic insect and plant species, found nowhere else. Spotted owls live there, and peregrine falcons. A quiet subspecies of red squirrel, found only on this mountain, has been declared endangered. A minimum of 40 mountain lions live there, and so do 150 black bears.

The mountain is not untouched. At least three sacred Indian sites have been found there, one by a bulldozer. Timber has been cut on its slopes since the turn of the century, and the Forest Service has cleared both roads and firebreaks near the top. Summer homes dot the road at lower elevations, and a lake draws campers and day-trippers. People come to hunt, to cut Christmas trees and firewood, to hike the trails, to ride dirt bikes. The Forest Service estimates that 200,000 people a year visit the mountain for one reason or another.

And yet, and yet . . . Being on top of Mount Graham is like being a thousand miles from civilization—or ten thousand years into the past. Here are hundreds of acres of untouched spruce-fir forest. The silence is overwhelming. Stop to catch your breath under an Engelmann spruce or a corkbark fir and you can hear your own blood thudding in your ears. I was there in early March when 3 feet of snow still lay on the ground. From the highway down in the desert, the mountain had looked the opposite of what I have seen in the Rockies. There is a tree line on Mount Graham, but here it is the line where trees begin partway up the slope and continue to the top. We drove up as far as we could in a four-wheel-drive truck and then switched to a treaded SnoCat belonging to the Arizona Game and Fish Department. For 7 miles we shattered the silence up a snow-drifted logging road.

Paul Pierce, leader of the 34-group Coalition for the Preservation of Mount Graham, had arranged the visit. Tom Waddell, an Arizona wildlife manager who has studied the mountain for 22 years, was our guide. A former rodeo rider, Waddell calls the proposed complex an "astrophysical Disneyland." He took us to a "show" squirrel, one who could be depended upon to be there. We floundered down a slope and sure enough, just above where Waddell had seen a shower of cone chips dropping out of a hole in a tree, an unremarkable squirrel poked its head out for a few seconds to look us over.

Opponents do not want the fight reduced to simply squirrels versus scopes, however. They talk about the cienagas, the unique wetlands above 10,000 feet. They talk about Apache trout and Apache goshawks. They talk about the bears being unscarred because conditions are so good they do not have to fight. Mostly they talk about a Pleistocene remnant unique in its biological mix, a treasure that could never be replaced.

The problem is that astronomers also consider the mountain unique, the University of Arizona's best chance to play a central role in the next stage of a golden age. These are boom years for astronomy. We are watching other solar systems form, finding out what galaxies are really like, and looking back in time closer and closer to the very beginning. At the same time, astronomers have come up with new designs that make it possible to build much bigger telescopes than were once thought feasible, at prices lower than anyone dreamed. The question is where to put them. The sites have to be as good as the telescopes. They need to be far from the cities, whose light is impairing existing telescopes. The University of Arizona, already a major player in the astronomical community, is pushing hard for the Mount Graham site.

J. T. Williams, the Mount Graham project director at the university, is there on loan from the Smithsonian. He spent several years on the mountain, testing its astronomical potential, getting around on a snowmobile, and says it is so good that "We get goose pimples thinking about it. The atmospheric transparency is just extraordinary."

The plans for Mount Graham are ambitious. The lineup of proposed instruments, as of April, included:

· The Columbus binocular telescope, an 11.3-meter instrument with more than twice the light-gathering power of any existing telescope
· A 10-meter submillimeter telescope for shortwave radio work, one of the few portions of the electromagnetic spectrum that is still largely unexplored.
· The 1.8-meter Vatican telescope. The mirror was built at the University of Arizona with the latest, spin-casting technology.

Four more are in the discussion-design stage, including an array of smaller, submillimeter telescopes that would work as one huge one, now being designed by the Smithsonian Astrophysical Observatory.

John Schaefer is president of Research Corporation in Tucson, a private foundation that funds basic research in all the sciences. He was once president of the University of Arizona, but he is also a founder of both the Tucson Audubon Society and the Arizona chapter of the Nature Conservancy. He believes that the observatory will be built on Mount Graham, and that it will save the mountain. Astronomers will take better care of it than will weekend recreationists on motorcycles, he argues. And the observatory could double as a biological field station.

At the other end of the power spectrum is Ned Powell, an Arizona spokesperson of Earth First!, himself a backyard astronomer. Some say his group is responsible for an unsuccessful attempt to smash the mirror of one of the test telescopes on the mountain. Powell says simply that Earth First! is not a formal organization, that two or three people may decide to do something, and may or may not let it be known that they are members of Earth First! His group is, prepared to appeal the Forest Service decision in August, but he points out that two Arizona Congressional delegates have said they will introduce legislation mandating the telescopes.

Walking on the snow under moss-draped trees, it was hard to imagine telescope domes, engineering shops, dormitories, fuel tanks, and all the rest. So the next day I drove down to Kitt Peak to see what an astronomical complex looks like. From the gallery of the 4-meter telescope there, the mountaintop looks like some kind of military installation, paved roads threading telescope domes, brick buildings, parking lots for heavy trucks, microwave and radio towers. Inside the visitor center are some pottery vessels found in a crevice on the mountain in 1960. A legend says they may have been an offering to I'itoi, the being the Papago people know as Elder Brother or simply Nature. The people's gifts restored the balance of nature for having taken wild foods.

The Forest Service originally recommended that the university be allowed to put five telescopes on one site, High Peak. In August, 1987, the university announced that they need a minimum of seven telescopes on two sites, High and Emerald peaks, to make the project economically feasible. This in turn required new studies of the potential impact on the now officially endangered Mount Graham red squirrel. The U.S. Fish and Wildlife Service was to finish that in May, 1988. Three months after that the Forest Service would decide.

The time for public comment, including mine, is over. The time for a disinterested panel of astronomers to determine what are in fact the best sites left in the United States is past. The situation is now analogous to a nonjury trial: Arguments have ended and the judge—the regional forester—has retired to chambers to reach a decision. Whichever way he decides, I will feel an acute sense of loss.

June 1988

[Note: By early 1993 the Vatican 1.8-meter telescope and the 10-meter submillimeter (radio) telescope were in their buildings and the buildings finished. They were expected to take "first light" in April or May. Construction of the twin 8.4-meter instrument, the Columbus binocular telescope, was to begin in 1994 with completion in 1997. Then, as required by law, the U.S. Fish and Wildlife Service was to again survey the red squirrel population before deciding whether any more telescopes could be built. The squirrel population had dropped slightly from the fall of 1991 to that of 1992, but was still twice what it had been in 1989. The Smithsonian Institution had decided to build its submillimeter array in Hawaii rather than Arizona.]

News Reports of a Dying Ocean

One of the fastest ways for a compulsive reader to get into hot water is to refuse to throw out old newspapers on the grounds they have not yet been read. Venial sins, like reading the cereal box at breakfast instead of participating in conversation, are quickly forgotten if not forgiven. A month of newspapers, stacked precariously on a kitchen counter, becomes a serious offense. Piles of older papers blocking traffic paths in the basement can lead to open hostility from housemates.

So on a Saturday when everyone was away and the temperature outside was on its way to passing 100 degrees with several to spare, I cleared off the dining room table, brought in the first pile, and went to work. Hog heaven is the technical term for my state when I confront a pile of newspapers and a free day. The hardest part is not deciding what to clip, but what to leave. It is amazing how significant a story can appear when you have time to appreciate it, instead of racing through the paper in the weekday morning rush.

With a clipper, scissors, and a straightedge, I plunge in. Something leaps out from almost every news section. In no time at all I have a fat folder full of miscellaneous intelligence, stories I want in my own primitive data base for some unidentified future use.

Some clippings are of conflicts I want to follow: a suburban couple who have turned what was their lawn into a meadow, running afoul of neighbors and the county authorities; Congressional efforts to impose some minimal safety standards on the nation's commercial fishermen, whose occupation is the most dangerous (and unregulated) in the country; and outrage in England and West Africa at plans to dump millions of tons of American waste there. Other stories are good news:

the birth of pups to red wolves reintroduced into North Carolina, or a meeting in Brazil on ways to cut hardwoods without destroying the Amazon rain forest.

One (perhaps the only) advantage of going through newspapers 50 pounds at a time is the chance to see patterns that might be missed from one day to the next. For example, no newspaper reader in my part of the country is unaware that Chesapeake Bay is in trouble. Harvests of oysters are way down, and the rockfish (striped bass) population dropped so low that Maryland imposed a moratorium on catching them. But there is a larger, national picture that is equally distressing, one that is harder to see unless you read the stories all at once. Here are just a few clips from that dining room table:

In southwest Florida, the number of young tarpon has dropped precipitously. A state team checked 40 places in the Florida Keys where young of the popular game fish have always been found: Tarpon were present in only five. Nobody knows if the drop is part of some natural cycle or an unnatural catastrophe. Russell Nelson of the Florida Marine Fisheries Commission is quoted: "The evidence leads us to suspect we have a problem with the tarpon, but in the absence of any funding for real research, it remains a puzzle."

In Albemarle-Pamlico Sound, the protected waters inside the Outer Banks of North Carolina, commercial fishermen are worried. First a fungal disease ate holes in crabs' shells and left gaping sores in the sides of fish. Then last year a brown tide—a bloom of toxic algae— hit the sound, forcing the closing of a hundred miles in the middle of the scallop and oyster season. Indigenous, nontoxic algae so deoxygenated the water that eels, crabs, and even fish were crawling out on the banks because they could not breathe in the water. This year the fungal disease is back.

On to the next pile of papers. In New Jersey, one story says, tens of thousands of dead eels, flounder, bass, crabs, and shrimp washed up on a beach inside Sandy Hook. State officials called the kill natural, the result of high temperatures and calm weather that left the water low in oxygen. Federal officials were less certain, however, because of the variety of species and sizes of the fish that died. By July the oxygen level in Long Island Sound was dropping toward the 1987 levels that killed lobsters and fish, and a brown tide had appeared in Long Island's Great South Bay.

Offshore the news is just as bad. Fishermen from Point Judith, Rhode Island, reported that the lobster catch from canyons along the continental shelf had dropped 70 percent in a little over a year, and of those that were caught, many were unsalable victims of "burn-spot" disease, holes in their shells eaten out by bacteria. Catches of tile fish, butterfish, and squid have dropped at least 50 percent. All this has happened since New York City and a number of New York and New Jersey municipalities began dumping their sewage sludge in 6,000 feet of water at the 106-Mile Site, so named because of its distance due east of Cape May, New Jersey. Until 1987, they had been dumping it in 80 feet of water just 12 miles off Sandy Hook, an area that has long been a "dead zone." The fishermen claim that some of the sewage vessels dump their loads before they get to 106-Mile, especially in bad weather, and that even the sludge properly dumped gets carried back up into the shelf canyons by eddies from the Gulf Stream. (Nine municipalities were threatened with fines totaling $1.25 million for dumping too much in one place, but none was accused of dumping short of the site.) Both the National Marine Fisheries Service and the Environmental Protection Agency (EPA) said there is no evidence that the dumping has anything to do with burned lobsters and the lack of fish, but the EPA is monitoring the area. (One solution has been around for 50 years: At the garden shop, I pay $8.99 for a 40-pound bag of Milwaukee sludge. The lawn loves it.)

Two general stories hung like a cloud bank over the rest. They concerned phenomena we have all read more than enough about, but which I, for one, had never applied to the oceans. The first said the Environmental Defense Fund had found that acid rain is doing direct damage to coastal waters. The nitrogen compounds involved are key nutrients leading to the algal blooms that deoxygenate the water and kill everything in it. The second pointed out that a depleted ozone layer and the increased "hard" ultraviolet radiation that comes with it could have a disastrous effect on marine phytoplankton, the microscopic plants that live in the surface layer of the oceans. Working in Antarctica, where ozone depletion is the most serious so far, Sayed Z. El-Sayed of Texas A&M University has found that any increase in ultraviolet radiation substantially decreases photosynthesis in the phytoplankton. This in turn would affect the entire food web that depends on them.

There's more, lots more, and two big piles of papers in the basement still to go through. But the pattern is clear enough now. Sitting at that table, fingers and elbows stained with ink, I saw the entire Atlantic shore like some magical museum diorama. The beaches were as beautiful as ever, the breezes just as fresh. But in the water, fish and shellfish were dying, and brown tides formed great, spreading stains. It was ugly, and led to further, frightening visions of the future.

The experts cannot yet say for sure exactly what is causing what. But the overall picture is there for a compulsive reader. Our coastal waters have taken all they can take and they're not going to take any more. Or, as one scientist put it, "We are exceeding the assimilative capacity of some of our coastal embayments."

The work that scientists are doing and will do is necessary but not sufficient. We do need to understand exactly what is happening, how it all works. But we already know, in a general way, what is going on. We have been treating the ocean as a waste tank and now, mirabile dictu, it is becoming one. We need to get angry enough to stop.

September 1988

Bad News from an Old Oil Spill

Suppose the whole world were wired, and each of us could sit at a giant control panel, like those at electric utilities and nuclear power plants, and watch what is happening to our planet. Facing acres of screens, lights, and dials, we would quickly become overwhelmed. The orange light on the central panel that flashed every time a species went extinct would quickly become part of the unseen background. We almost certainly would not notice screens that had gone blank, lights that were no longer lit, even if we had been briefed that no news could be bad news. We might be like the cub reporter of legend who was sent out to cover a town board meeting and returned to report that there was no story because the board had not met. Only when the city editor asked why did the novice mention that the town hall had burned down.

Or imagine a miner alarmed that his canary has died. He returns to the surface only to be told that all the canaries in their cages in houses around town have died, too, and that therefore there is nothing wrong down in the mine. He could be excused for thinking that something is wrong everywhere.

Large oil spills in the ocean light up our environmental instrument panels. The *Amoco Cadiz*, the *Argo Merchant*, the *Exxon Valdez* all made the front pages of our newspapers. The worst spill of all time occurred during the Gulf War. Yet we often do not learn for years, if ever, just how much damage to the ecosystems was really done. This happens because our knowledge of what was there before the spill is almost always incomplete, and because we do not know what other harmful forces are simultaneously at work. Some of the biological studies are of the unrefereed kind known as "gray literature," and some are kept secret pending the outcome of lawsuits. We see lots of pictures of oiled birds and sea otters, but we never quite get the big picture.

Thus for ecologists one of the most interesting spills of recent years was one that occurred on the coast of Panama 5 years ago. On April 27, 1986, a storage tank at a refinery on Pavardi Island failed, spilling more than 50,000 barrels of medium-weight crude oil into the Caribbean. For the first six days, onshore winds pushed the oil into an adjacent bay. On May 4 rainfall and shifting winds pushed the oil out to sea, where it could not be controlled. In the following weeks more on-shore winds distributed the oil along the coast and into bays and estuaries.

The spill occurred about 3 miles east of the Galeta Marine Laboratory of the Smithsonian Tropical Research Institute (STRI), which in turn is only about 3 miles east of the Caribbean entrance to the Panama Canal. What made this spill unique was that STRI biologists and visiting scientists had been studying the reefs, algal flats, seagrass beds, mangrove forests, estuaries, and sand beaches on just this stretch of coast for 15 years before the spill. They were there when the oil came ashore, and they have been doing meticulous studies ever since. What makes it news is that their findings to date contradict much of what we thought we knew about spills. In a review of the situation published in *Science* in 1989, the 18-person team concluded:

> . . . other results were not expected, and in some cases contradict widely held views about the effects of oil spills and the ways they are studied. First, extensive mortality of subtidal corals and infauna of seagrasses had not been demonstrated before and contradicts undocumented assertions that these organisms are not affected by oil spills. . . . Second, the magnitude of subtidal coral mortality and injury . . . is in striking contrast to findings of no lasting change in coral condition or growth after exposure to oil . . . in small-scale experiments. . . . Third, sublethal effects are extensive and may be more important in the long term than initial mortality.

The severity of the damage is all the more remarkable, they point out, because the area has long been under environmental assault:

> Much of Bahia las Minas has been subjected to human disturbance, beginning with decades of excavation, dredging, and landfilling for the construction of the Panama Canal and the City of Colon, drainage and spraying of mangroves for mosquito control, construction of the refinery and a large cement plant on landfill, a major oil spill in 1968 [the wreck of the *Witwater*], and unknown amounts of chronic oil pollution from the refinery and ships passing to and from the Canal. It is a measure of the severity of the 1986 oil spill that the biological consequences were so detectable despite this history of environmental abuse.

Last April I visited Galeta with Brian Keller, manager of the oil spill project. We drove from the Pacific to the Caribbean, onto the grounds of the U.S. Naval Security Activity Group at Galeta (the Navy provided the laboratory building for STRI). I saw biologists working a transect on a reef flat at extreme low tide, carefully censusing the sea urchins in square-meter plots. A boat brought divers back from their morning's work. No oil was to be seen along the reef, in the wrack at the water's edge, or in the mangroves we walked along. Egrets were fishing in the mangrove channels; a shark cruised the shallow bay next to the station. If someone hadn't told me, I wouldn't have known that anything was wrong.

Keller told me that about 180 acres of mangroves had been killed, mostly in strips along the water's edge. They will probably recover in 20 to 30 years. But 5 years after the spill the reefs show no sign of recovery. Atlantic reefs reproduce slowly even in the best of circumstances, he said; coral heads grow around 5 millimeters a year. On the Panama coast, even outside the area of the oil spill, they are doing poorly. Before the spill a good reef was about 30 percent living coral. Those adjoining the spill site have only about 10 percent. So there is little chance that the nearly bare limestone inside the spill area will be recolonized anytime soon. This failure to recover is not simply a direct result of the spill but of the widespread decline of Caribbean reefs. All the canaries are dying.

The contract for the oil spill study, from the Minerals Management Service of the Department of the Interior, expires next March. Right now there is no funding to make this a truly long-range study. But nothing less will do, it seems to me. A government report issued last April, for example, on the damage to Alaska's Prince William Sound from the *Exxon Valdez* spill 2 years earlier said that far more wildlife had been killed and the environmental damage would last much longer than scientists had originally thought. Some of the damage will be permanent. We should go back in 5, 10, and even 50 years so we will really know.

To call for better understanding of just how damaging oil spills really are is not to declare open season on oil companies. We may indeed need to do more with dikes around oil tanks and double bottoms for tankers, even if the latter not only raise the cost of ships 15 percent but also present their own problems. Shipowners argue that when a double-hull tanker runs aground, explosive fumes might accumulate in the space between the hulls. Or water might accumulate there, causing the ship to sink lower into the sea and leak more oil or even, in the worst-case scenario, capsize. Legislation passed by Congress last year requires double hulls by 2010, however; several of the major oil companies are now building such ships.

As long as Americans choose to consume 17 million barrels of oil a day, mostly by burning it, tankers will sail the seas and mistakes will happen. (Our descendants may shake their heads at the thought of us simply burning the stuff when it is far more valuable as feedstock for the plastics and miracle materials we so dote on.) From the house I grew up in on the island of Aruba, you could see an oil tank farm to the north, the refinery to the west, and an anchorage to the south. Children swimming in the lagoon would stop to watch as an oceangoing tanker came around the point and ghosted west just outside the reef, its great curved bow high out of the water, the rapidity of its flashing-light query for berthing instructions setting off the enormous power of its kinetic energy palpable even at dead slow. We traveled by tanker then, meeting sailors who had had three ships torpedoed out from under them, who knew all about being oiled. Even in peacetime, large ships carrying toxic, inflammable cargoes are inherently dangerous. Tankermen are careful for the same reason airline pilots are.

Oil tankers will sail the seas, in good weather and bad, breaking apart in storms, catching fire, running aground, as long as we keep burning the stuff. My favorite antiautomobile advocacy ad after the *Valdez* grounding was the one that featured a picture of the captain and a headline that ran: "It was not his driving that caused the spill; it was ours." I wonder how complacent we would be if there were a light for each of us on that mythical control panel, whose color indicated our personal impact on Earth.

August 1991

Home to Aruba—Too Late

For years I have had a fantasy about returning places to the wild, taking down houses and taking up streets, letting Nature come back at her own pace and in her own way, so that in 10 years or a hundred or a thousand no one could ever tell we humans had once been there manufacturing, consuming, amusing ourselves.

In all those hours of daydreaming, often about Manhattan as a new Eden, it never occurred to me that they would start with my house, my street, what was the very center of the Universe when I was discovering variable stars, bird guides, darkroom magic, and the far more magical fact that girls are indeed different from boys. But turn my back for a trifling 38 years and wham! Look what they've done. Bungalow No. 415 is gone. The street it was on is gone. So is half the town. Crested caracaras now stalk through the cactus where children once played on manicured patios.

You come into this particular town from the north. After seven or eight blocks, in my day, you came to the top of a hill with a church on your left and the lagoon in front of you. Down the hill is the beach. We are not talking Los Angeles here. So my friend and I came flying into town in our rusty rental car one day last March. We pulled to a halt at the first stop sign. And there, across the street on our left, was the church. Straight ahead was the lagoon. We had missed most of the town because it no longer exists.

The rabbits are reclaiming what was theirs. Iguanas scuttle into the thornbushes, parakeets scream like green rockets through the organpipe cactus. I should be rejoicing. Part of me is. But where are the school, the commissary, the ball field? Where is my youth?

This once was Lago Colony on the Dutch island of Aruba. The

refinery— and the company town—were owned by the Lago Oil and Transport Company, eventually a subsidiary of what is now Exxon. Once the refinery was one of the largest in the world, and the port of Sint Nicholaas one of the busiest. But in 1985 the refinery closed for good. Now Lago, or Exxon, would appear to be in the restoration-ecology business. Not only are the streets being taken up, but a sign at the entrance to the colony announces that the whole place is a sanctuary and neither animals nor plants are to be disturbed.

Some traces of a boy's world remain. The last house we lived in is still there, windows boarded up, yard overgrown. A tree now covers the garage from whose roof I watched the stars; hummingbirds have built a nest in the bottom branch of another tree that now blocks the back door. Other pieces still stand. The Esso Club is still down by the water, on a point between two lagoons. The tennis courts are still there, and a snack bar on the beach where the "yacht club" once stood. The golf club is still going strong, sold to its members for one guilder. But it was monumentally disorienting to drive around a small, familiar town, looking for friends' houses and other childhood landmarks, and not be able to find them because they have ceased to exist.

It was discombobulating enough for me that we spent the next three days away from Lago, driving around the rest of Aruba. A 19-mile-long volcanic island encrusted with laminations of long-dead coral reefs, it lies athwart the northeast trades. On the windward side, the surf blasts holes in the coral cliffs; on the other, mangrove-lined lagoons doze under the painful sun. Indians crossed the 20 miles from Venezuela 4,500 years ago and gathered shellfish in leeward lagoons while, as a brochure from Aruba's archaeological museum puts it, "at the windward side some real fishermen lived." There is a lot to see, despite the Denver Post travel writer who called Aruba "the world's ugliest island. This place is so devoid of anything to do," Michael Carlton went on, "that even Gilligan would flee." Luckily I knew that already, so I was able to ignore the wind surfers jumping the breakers at Boca Grandi, golfers staring down burrowing owls in the sand traps at the Aruba Golf Club, snorkelers mixing it up with French angelfish and sergeant majors in the staghorn coral off Baby Lagoon, the sailors anchoring off Malmok for a picnic on a miles-long beach, binocular-laden eccentrics counting wood storks at Bubali or parakeets in Frenchman's Pass, time travelers pondering birds and fish the Indians had painted on the rocks in what is now Arikok National Park.

Because there was nothing to do, my friend and I walked through the dunes to the cliffs of California Point, past the stone huts of today's real fishermen and the crosses cemented to the rocks in memory of the drowned. We drove between the edge of the cliffs and enormous piles of broken rock down the coast to Alto Vista, where the first Catholic mission meant the beginning of the end for the Indians. We drove out through Santa Cruz and the national park to the coast, to find a triple natural bridge above the surf, and back through Andicuri, a coconut palm plantation thought to carry the name of an Indian chieftain. We climbed the crumbling concrete steps of Hooiberg ("the haystack," a conical hill at the center of the island) and the red rock of Seroe Colorado, the lighthouse hill at the southeastern end that looks down on what is left of Lago.

At Boca Prins, in a valley behind a break in the cliffs where a boy camped 40 years ago, we met an American, ecstatic that he had left his asthma back in Washington; and an Aruban casino dealer, ecstatic at the prospect of his upcoming trip to Las Vegas. Another Aruban, just back from the Netherlands, had spent the day on the cliff above us staring at his indigo sea. We met more Arubans, eager to help in four languages, while looking in Siribana for the house of a classmate from long ago. (We finally did find her house, but not her, in another part of the island called Shiribana—I had forgotten the h.)

On the south side we drove the beaches, walked the mangroves, swam in green water on pink sand. We drank beer in the rum shops, and ate beef croquettes and chicken pastries in the bakeries of Sint Nicolaas. We ordered nasi goreng (Indonesian fried rice) one day in Oranjestad as frigate birds soared over the boats that bring fresh produce from Venezuela; later we dined on fresh fish (caught from the boat tied up next to our table) as the sun set over Spanish Lagoon. We tried the goat stew in an Aruban house, and sipped iguana soup and rum beneath the watapana and kwihi trees while the yellow satin and white linen of the Aruba Dance Theater swirled under the stars. Then back to the hotel, huge clouds sailing right over our heads, stars burning in the gaps, the warm, soft wind almost itself alive, bringing bird noises from the wetlands sanctuary next door.

By the end of the week we were back in Lago, swimming in the lagoons, having a beer in the bowling alley at the Esso Club, even playing a round of golf for old times' sake. I walked the aboveground sewer pipe along the shore of the lagoon, looking for the iguanas that used to leap into the water and hide under the rocks; those I found headed uphill instead.

"One Happy Island," the license plates say, and it really seems to be, remarkably so. Physically, however, I must confess to my Colorado colleague, it is not quite paradise. Things are tough in the best of times on this island, and the past 500 years have not been the best of times. Average rainfall is only about 17 inches, usually in quick downpours from October to December. Plants that do get a start face the salt-laden trade winds. The first Europeans cut down the dyewood trees, and introduced the goats and sheep that are causing murderous soil erosion to this day. Although it is hard to believe, looking at today's landscape of cactus and bare dirt, the entire island was once used as a horse-raising station. While still under Spanish rule, the island was a source of Indians for slave labor in the mines of Hispaniola. Gold was discovered in the 19th century; parts of the island were ripped up in the frenzy. Now, stone quarries and sand and gravel pits scar the land everywhere as the hotel building boom goes on.

A new fantasy takes shape. It is night. A small, twin-engine plane flies low over the sea, without lights. It lands at the old Aruba Flying Club, whose X-painted runway now heads straight for the new prison. A man jumps out of the plane with a rucksack full of old coffee cans. He runs off into the cactus and thornbushes, drops to his knees and tears open the pack. He pulls the lids off the cans, dumps their contents on the ground. Then he runs, crouched, back to the aircraft; the plane leaps off the runway and up into the wind, and disappears back over the sea. Police cars come screaming up; cursing men, with flash-lights, fan out into the brush. One yells and the others converge: there in the beams of their lights is a dark mass of wet topsoil, compost, and rich loam from an old vegetable garden. The phantom island restorer has struck again.

May 1989

Disturbance Is Natural, Too

 As a sometime weekend naturalist, I like to get out into nature whenever I can. Yet, for someone living in the East, nature is almost never natural. It is unlikely that I have ever walked on ground that has never been plowed or through a forest that has never been cut. Mammalian food webs have been amputated at the top: Cougars and wolves are long gone. Really big trees are as rare as big animals. There are precious few places to see the mature forest the colonists saw, never mind the truly primordial forest the first arrivals found.

Instead we have the intermediate stages of a forest that has been disturbed, what biologists call the stages of succession. Species succeed species as the disturbance fades into history. Human impact is visible everywhere. Bits of rusted barbed wire and the remains of old stone walls materialize in the middle of a seemingly pristine wood. Odd bits of engineering—an old culvert, the piers of an abandoned bridge—appear along a stream. One pushes off well-worn paths, only to find more beer cans, all the while bathed in the acoustic waves of jets climbing away from the airport.

Disruption is the norm. Where to find nature undisturbed? The question, I'm only just now finding out, is naive. The natural world is constantly being disturbed. From the fall of a single tree to devastation by hurricane or forest fire, that world is changing all the time. What I had not understood is that the natural world does not recover from such disturbances the way a body heals from a wound; ecosystems depend on disturbances, indeed could not live without them. Recently I ordered a book because I had misread the title. It is *The Ecology of Natural Disturbance and Patch Dynamics*, edited by S. T. A. Pickett of Rutgers University and P. S. White of Great Smoky Mountains National Park. I had overlooked the word "natural" in the title and thought the book would be about the cutover forests and plowed fields where I spend my time. But the papers in this collection are indeed about natural disturbances, and for me they opened up a new understanding.

The central idea is that a seemingly homogeneous habitat, a particular section of forest or grassland, is really a shifting mosaic of patches, sometimes very small fractions of the total, where a disturbance of some kind has created a difference. The concept arose at least 50 years ago, in studies of tropical forests. When a tree falls, it creates a gap in the canopy. The gap creates a patch, a small area where the environment is different from the surrounding forest. The most obvious difference is more light. Tree species that could not grow, from seed or sapling, in the subdued light now have a chance. These species are an integral part of the community, may even be a major component, yet can only take their place when disturbance makes it possible. A tropical forest is a dynamic system of such patches.

The grasslands of our continental interior are another case in point. Traditionally the prairies have been thought of as climax communities, in which succession has led to the final, stable composition of plant species, so even fire has little or no effect on the mix of species present. In the new view a prairie is a sea of small patches, most often caused by small disturbances, that help produce the species mix we see. Orie L. Loucks of Butler University, Mary L. Plumb-Mentjes of the U.S. Army Corps of Engineers, and Deborah Rogers of the Technical Information Project in Pierre, South Dakota, who have studied the Spring Green sand prairie in Wisconsin, explain the idea. The patches include the work of pocket gophers and badgers and, on an even smaller scale, include mounds built by ants. Certain species of opportunistic plants, major constituents of the prairie, often rely on such openings. Other papers treat the importance of disturbance and the patterns of patches in such ecosystems as shrublands and rocky intertidal zones, for such groups as insects and vertebrates, and for energy and nutrient pathways. Still others spell out the implications for genetics and evolution. A good deal of it, I must confess, was beyond me. But the book gave me a new way of looking at things.

During the winter I had spent much time along the Potomac River northwest of Washington, wandering through the sycamore-dotted alluvial sands and across the rocky outcrops of the fall line, around the little ponds and across the streams that lie between the river and the Chesapeake and Ohio Canal. The whole strip looks like it has been bombed. Heavy logs are nestled in the branches of small trees. Entire stands of trees have been knocked over. In places the sandy soil has been swept away, leaving tree roots exposed 3 feet down; in others, mounds of sand and polished, rounded river rocks have covered what had been luxuriant plant life. A spent hurricane had dumped its rain on the Potomac watershed the previous November, producing the worst flood in 13 years. This is a disturbance of the first magnitude, and it is all completely natural.

At first, being there was as depressing as walking through a woods after it has first been lumbered and then stripped for firewood. But after reading Pickett and White, I thought some more. This stretch of river is no gentle floodplain, but it certainly floods. The worst in this century occurred in 1936. From then it was nearly 40 years to Hurricane Agnes in 1972, and another 13 to the latest. The period may be irregular, but flooding is certain. How then to look at that vulnerable ribbon of land between canal and river?

It is time to adjust my thinking and shorten my gaze. What strikes my eye as a mess, even ugly, is the real world, the natural world, and I must learn to see it that way. To do it I have to consider each patch in and of itself. What will come up where the trees are down? Will the sand piles spread out in the spring rains or be colonized where they are?

Doing so may teach me something about disturbances and something about patches. These concepts are not new, but neither are they widely invoked by biologists. In a paper on disturbance and vertebrates, James R. Karr, then at the Smithsonian Tropical Research Institute, and Kathryn E. Freemark, of Carleton University in Ottawa, offer an explanation of why, until now, most work in ecology has concentrated on the "presumed equilibrium nature of biological systems." Even recent mathematical approaches to ecology have required assumptions of equilibrium in order to be easily handled. Second, ecologists often make definitive statements from short-term studies, periods that may not be representative of the organism or biological system in question. Or they may average data from several study areas or from several years, so that real differences are inadvertently masked. These problems arise from the rules of the game—short-term research grants, the career need to obtain and publish results quickly—that discourage the long-term studies required to understand the role of disturbances and the resulting heterogeneity in biological systems.

Even a biologist who works hard at seeing the fragments of an ecosystem may still not be home free. In what for me was the best paragraph in the book, John A. Wiens of the University of New Mexico ends his paper on arid and semiarid ecosystems with a caution. The real trick, he argues, is to comprehend the patches in the same way that they are perceived by the organism being studied. The problem is not only that we see the environment "on a scale different from that of an aphid or an ant" but that we tend to emphasize whatever is most important to our dominant senses—especially vision. "Other organisms," he says, "may perceive environmental mosaics in quite different ways." Unless we can understand and adopt the organism's view of the environment, we are unlikely to discern what is really important and instead produce "patterns that are little more than artifacts, products of our misperception of reality." The correct answers, Wiens says, will come from study of the natural history of an organism and especially the changes in its behavior as it exists from place to place.

My work is cut out for me as I stumble over trees and slip in the sand. A hurricane has given me a prodigious disturbance to examine. It is up to me to find the patches, both those I can see and those the inhabitants see. And perhaps as the years go by I will begin to understand what the disturbance has done for, rather than to, one strip of land along a river.

March 1986

PART 3

Wildlife Is Where You Find It

The good news, sometimes lost in all the alarms from the front, is that nature is damned hard to kill. Life is everywhere around us, probing for a place to live like the wind-blown and sea-borne arrivals on a newly formed volcanic island. In the black cinders of railroad beds, *Ailanthus* (the Chinese tree of heaven) unfolds its pinnate leaves. A cottonwood manages to take root and grow atop one of the towers of the Queensboro bridge linking Manhattan and Long Island. In the meanest vacant lot, arthropods live their complicated lives in a jungle of opportunistic plants (weeds to the unknowing).

Nature will meet us more than halfway. Plant a bush, dig a goldfish pond, put up a birdhouse—then stand back out of the way. On a short bank along the driveway of a home I once owned was a rock garden that I simply ignored. In 5 years that unpromising patch of dirt became a miniforest of red maple and oak; one tulip tree was 15 feet tall and growing fast. The birds and an army of invertebrates had long since found it, of course.

Because nature is everywhere, we do not have to drive to the local nature preserve to find it. It is in our gardens, by the side of the road, in drainage ditches, in the bushes in front of the bank. As Edward O. Wilson wrote in *Biophilia*, "To the extent that each person can feel like a naturalist, the old excitement of the untrammeled world will be regained. I offer this as a formula of reenchantment to invigorate poetry and myth: mysterious and little known organisms live within walking distance of where you sit. Splendor awaits in minute proportions."

The tale is told of an entomologist at the American Museum of Natural History in New York who lobbied for an exhibit on the insects a homeowner might expect to encounter. Not enough variety, he was told. The curator persisted: How many species would he have to find in his own backyard in a typical New Jersey subdivision to convince his superiors? Seven hundred was the ruling. The entomologist went away, and several months later returned with his documented findings: 1,500. He got the exhibit.

Occasionally the situation is reversed, and a piece of the city is transplanted into a wild place. What could be more urban than the sight of a tower crane swinging buckets of concrete to the top floor of the latest high-rise? Inspired scientists placed one in a Panama forest, so buckets of biologists could study the hitherto inaccessible canopy as easily as birds.

We don't need a tower crane in a tropical forest to find the natural world, however. It is everywhere. All we need do is look.

Make Room and They Will Come

 It always surprises me to walk into a brand-new restaurant and find it packed. Presumably all those people had managed to eat the day before. But there is a kind of life pressure that rushes into any new opening, fills any new space, like water pouring over uneven terrain.

People produce the most tangible life pressure. Roads, houses, shopping centers, factories spill over natural habitats like an onrushing tide. Humans push the wild into smaller and smaller spaces, leaving only the occasional pocket as park or refuge. One envisions wildlife confined to islands in a sea of human development.

But this dominant, human pressure is not the only life force rushing in to fill openings. Lots of traffic moves the other way. The wild is all around us, and pushes in on us whenever and wherever it can. The wild is searching, circling, watching, waiting for a new opening even more intently than city dwellers look for new restaurants.

Like many a city dweller, I tend to head for those parks and refuges when I want a taste of the wild. But all I really have to do is raise my eyes. There have been summer evenings when I dragged home from a day in the marsh, only to find more action in the skies above the city rooftops than I had seen all day. The nighthawks had come out and were swooping through the dusky light after the insects I could not see. And there was a morning in a quintessentially urban environment, standing on a subway platform (at an aboveground station) in a crowd of commuters. I looked up from my paper to see what the hammering was and saw a pileated woodpecker, as big as a crow, all pterodactylous black triangles, red crest flashing as it worked an old tree in back of a caterer's establishment.

The wild sneaks in every time we blink. I walk down a wide concrete sidewalk between an eight-lane avenue and a row of concrete-and-glass government buildings. There is a government gardener, meticulously pouring poison into the expansion joints in the sidewalk, trying to kill the grasses that have taken root there. She may win the battle, but she will never win the war.

This life pressure can be seen best by deliberately creating new openings. A happy band of subversives in the National Park Service (NPS) has been doing just that here in Washington, with extraordinary results. They have taken a couple of architect's reflecting pools and converted them from algae-ridden trash pits to living ecosystems.

It all started in Bolivar Park, one of those little triangles that grace the city. The central feature is a concrete reflecting pool, set squarely in front of the Department of the Interior building across the street. As John Hoke of NPS relates the tale, the question arose as to whether there should not be a little life in a pond in front of the building that is headquarters for both the Park Service and the Fish and Wildlife Service. The pond already cost money to manage: It had to be drained several times a year to get rid of algae and litter on the bottom.

After an extended discussion of the esthetics involved, Hoke and his colleagues—in the NPS Division of Resource Management and Visitor Protection and in the Ecological Services Lab—got the go-ahead. They planted cattails and water irises in boxes of swamp soil just under the water surface. They introduced enough floating vegetation to cover about a third of the pond. They put in a large log and then turtles to climb out and bask on it. And best of all, they built an island, no more than six feet across.

A postage-stamp island in a concrete pool next to a busy intersection in a fair-size city would not have appeared very promising to me, but then perhaps my own life pressure is a little sluggish. Ducks had flown far enough into the city to spot the island, and then quickly made it their own. Every year one or more broods of ducklings are hatched there, feeding on the aquatic vegetation (and some judiciously spread trout meal) until they are ready to travel.

In the spring, office workers from Interior come out at lunchtime to

watch drakes battle for breeding rights while resident hens band together to keep interloping hens out of the pool. The crowds of people
are heavy enough so that the Park Service has had to replace the
original lawn around the pool with a tougher fescue and downgrade
its maintenance standards. Yet hardly any trash is thrown into the
water anymore.

The same magic has been worked in the seven-acre pond in Constitution Gardens along the Mall. Designed and built strictly as a formal
reflecting surface, the lake was seeded with submerged vegetation in
underwater planters; the one tiny, wild island was immediately com-
mandeered by breeding ducks.

Along the way, Hoke and his colleagues have found a secret ingre-
dient that keeps their ponds in balance. They were still having trouble
with algae, even after they resorted to an inert black dye that darkened
the water enough to slow algal growth. So last July, when a particularly noxious bloom occurred, the Park Service took a pumper truck
(the kind used to empty out septic tanks) to Kenilworth Aquatic
Gardens, loaded up 500 gallons of bottom muck and inoculated Bolivar Pond with it, much like adding sour-dough starter mix to a batch
of pancake batter. In weeks, algal growth stopped.

The same miracle muck has been added to the lake in Constitution
Gardens and the Reflecting Pool that lies between the Lincoln Memorial and Washington Monument, with the same happy results. The
accountants are happy, too. Where it had cost $2,000 to drain Bolivar
Pond to get rid of the algae (sometimes that had to be done four times
a season), it had cost $20,000 to drain the Reflecting Pool.

Hoke and his colleagues have a history of working with nature
instead of trying to overcome it. Years ago, when a cluster of "tempor-
ary" Navy building were removed from the Mall, some Park Service
managers were dismayed to find a low, soggy wet area that quickly
became a depository for beverage cans and windblown newspapers—
and a breeding ground for mosquitoes. It made no sense to install
expensive drains, for the area was to be torn up again for the creation
of Constitution Gardens.

So they went the other way. The area was trying to be a wetland:

They would help it along. They scooped out a hole deep enough for fish life. Next they raided a natural swamp for cattails and water irises, water lettuce and duckweed, mosquito fish and bullfrogs. Dragonflies and water striders arrived on their own, as did, naturally, flotillas of wild ducks. For the price of a few hours work with a backhoe and one truck trip to a swamp, they turned a seep into a flourishing wetland that quickly became popular with noontime picnickers. That pond and swamp are long gone, but the philosophy behind them is still very much alive.

At the entrance to the Park Service's National Capital Region headquarters on Hains Point is a concrete pond where experimenters can try out new combinations of plants and pond organisms. Bass and bluegills live there, and it is not unknown during the lunch hour for a Park Service worker to step outside with a fly rod, let out some line in a few false casts, and drop a tempting tidbit next to a lily pad.

Bouncing along the Mall in his electric cart, Hoke laughs with pleasure at how fast life moves into every new space he manages to create. He wants to know more about why what he does works so well. He hopes that somewhere a graduate student will take on the mission of finding out exactly which herbivores and decomposers in that pond muck—which microorganisms, aquatic worms, and insect larvae—play the key roles.

He thinks about other wild creatures that circle our city islands, always looking for a new opening. Not long ago he was in London, trading information with the people who manage the Serpentine Lake in Hyde Park. As many as 50 species of unusual waterfowl are seen there in a year, he was told. And it is not clear to him why we cannot do better here. If John Hoke has his way, there will be a lot more traffic coming the other way.

February 1984

Why Not Rooftop Refuges?

 Imagine thousands of acres of clear, level surface without a thing growing on them, baking in the sun, cracking in the cold, scoured by the wind like the landscape of some lifeless planet. Every city, every industrial park has such places, usable space from the human point of view because it occurs where humans are most concentrated.

The space in question is that on the roofs of our large buildings. Those acres and acres of tar and plastic are broken only by outcroppings of elevator shafts, vent pipes, and air-conditioning machinery. And yet they could be the substrate for green, growing things, the foundations for gardens in the sky. The idea is not new. People have been growing plants on their roofs at least since the time of Babylon. But the practice is not widespread. It is the exception rather than the rule, more associated with the penthouses of the wealthy than the workaday world of office and factory. And yet putting a garden on the roof is good for a building. As we shall see, it helps eliminate the flaw shared by nearly all large buildings (no matter how carefully they were constructed): A roof that leaks. It also turns wasted space into valuable real estate.

Such gardens not only are good for the individual buildings they sit on, but improve the quality of life all around them. They are good for people and, even in the middle of the largest city, good for wildlife. Planted roofs create refuges for people—office and factory workers on their lunch hour, apartment dwellers any time. And they can be refuges for animals. Migrating birds gain a place to rest and feed, and certain endangered species could be protected there. What safer place for an endangered turtle, proponents argue, than a pond 10 stories above a city street, where most predators cannot reach it?

Roofs are the single most intractable problem in large buildings; no matter what you do, they leak. Conventional roofs, built up of layers of tar and felt, eventually leak, so the owners put on more layers until finally they have to scrape them all off and start again. Newer technology uses plastic membranes that do better than tar, but in time they, too, will leak. A rooftop is a tough place to be. Temperatures can vary enormously from noon to midnight. Moisture levels go from standing water to bone dry. Winds carry dirt and grit, abrading the surface. No covering can last for long.

Tar, after all, is a mixture of substances; lots of volatile hydrocarbons "boil off" as it cures, leaving pores and cracks where water can enter. The new plastics are better, but they are hard to put down perfectly (no roof is completely flat; instead, it will have high and low spots) and hard to seal tightly against all the apparatus that sticks up through a roof. (These last expand and contract at different rates as the roof heats and cools.) And plastic degrades in the sunlight that beats down on it every day.

Dirt turns out to be a miracle roofing compound. Its natural milieu is exactly that harsh environment that wreaks such havoc with roofs. Consider: Dirt is composed of small particles; it can flow. It moves when vent pipes or other obstructions expand and contract, keeping tight contact. Dirt dampens the swings in temperature that crack roof materials. Farmers well know how long dirt takes to warm up. And dirt reduces oscillations between wet and dry, maintaining a more constant moisture. A layer of dirt protects the roof itself from wind abrasion and keeps sunlight from reaching any plastic membrane underneath.

James W. Hudson is an apostle of rooftop planting. A chemist by training and an OSS agent during World War II, he is now a design consultant and engineer. He has always been innovative, coming up with ideas like perforated beams and joists that carry electric and telephone wires without sacrificing any strength. It is he who so eloquently reels off the advantages of dirt on the roof. As he talks, he sketches different types of roof construction and argues that almost any building can support some kind of garden on top.

For about $1,000, he says, engineers can determine the load-bearing capability of a roof. The inspection is absolutely necessary, he feels. In theory one could go to the original design drawings, but in practice the specifications may not have been followed. When competent engineers look at the structure from underneath, he says, no guesswork as to what the roof can stand is involved. (Hudson is adamant that no one under any circumstances should put extra weight on a roof without calling in the engineers. "Don't do it yourself" is the first rule.)

With proper precautions, Hudson says, almost any building can take planting on the roof. The design strength of many buildings, based on the average expected snowfall, is about 30 pounds per square foot. But the columns supporting the roof can bear thousands of pounds. So if need be, the first step is to build a planter that spans all or part of the roof. Particularly heavy items, like trees, can be planted near the columns.

Planting a roof is not just possible in theory. Decorative gardens adorn the roofs of buildings all over the country, like the five-story parking garage at the Kaiser Center in Oakland, California. In New York City a warehouse roof has been turned into a commercial hydroponic garden. And plantings can be as elaborate as desired: One rooftop garden in London has thousands of species growing in it.

Over and above their value to the buildings involved, roof gardens can be pleasant parks and even playgrounds for the people who work and live in them and a pleasant sight for those in surrounding buildings. They benefit the whole area. By their mere presence, plants act as passive traps for dust and soot. Plants in leaf have a tremendous surface area, which makes them better collectors than, say, a vent pipe. Their unique contribution is photosynthesis, in which plants take in carbon dioxide and give off oxygen. Rooftop gardens could do for city air, in a small way, what bubblers do for an aquarium.

And they could be refuges. Birds certainly will come. Canada geese have nested on the bare 10th-floor balcony of a bank building in St. Louis. And a recent experience at Rutgers University, while not involving a roof garden, may be instructive. On a very urban campus in

Newark, on a small plot surrounded by concrete buildings and a parking lot, there stands a greenhouse. Some of the area outside is wild, some cultivated. According to a report in *American Birds*, both areas are being used by birds. Wood thrushes and ovenbirds have been seen there, as well as catbirds and a number of others.

Imagine if that plot had been on a roof, so that the birds did not have to come down below the level of the adjoining buildings to get there. One could reasonably expect that even more birds would have dropped in. In fact, imagine what a city looks like to a bird flying overhead. Mostly the bird would see a totally alien landscape of tar and plastic surfaces, sterile stretches with no cover, no food, no water. For it, the city is only a place to be flown over. But suppose the same bird looked down to see islands of green bushes and shrubs, even small pools and waterfalls, rising up to invite it. Now the bird has the option of stopping. At least some will move in at the slightest opportunity.

The greening of cities seems to me to be an unequivocal good. I have always thought that certain things in life are virtuous per se. Buying a book is one; it is tangible support for the life of the mind. Planting green things is another; it always improves our environment. Something in us needs at least a touch of the natural world. We do what we can. We may start with a few houseplants and go on to a wall of living green. We may attach a window box to the outside or fill a terrace with potted plants. A backyard can be transformed from a sterile expanse of struggling sod into a garden that offers endless refreshment for the eye. For those so inclined, the National Wildlife Federation can help turn a backyard into a pocket refuge.

Even cities, the seeming antitheses of landscape, are softened by greenery. The hard surfaces, the straight lines and right angles are less harsh when their geometry is broken by plants reaching for the light. A few trees along a sidewalk, a vest-pocket park in the middle of a block can work wonders. Higher up, any place where there is a balcony or a ledge, humans instinctively try to grow things. Inside, in their apartments and offices. And outside, anywhere they can put a pot or a tub. Some cities are famous for such efforts, cities as disparate as

Rome and London. But the effort goes on everywhere, and all cities are the better for it.

Hudson and his fellow enthusiasts want to go further, to plant entire roofs, to turn whole cities into gardens in the sky. His good news is not only that it can be done, but that it is a practical thing to do. I try to imagine what it would be like to gaze out of an office window in some high building, and look across and down at a succession of gardens, parks, and refuges, a steplike sequence of green places thrusting up to the light. I like it a lot. I say we go for it.

March 1985

Flying Fertilizer Factories

Sometimes the hardest advice to follow is: "Don't just do something! Stand there!" By nature, it seems, we humans are interveners. We manage our sanctuaries and preserves for some favorite bird or animal. Rather than putting up signs and leaving the place alone, we send the engineers into wildlife reserves to build miles of dikes and dams and drainage ditches. We want to do it ourselves.

Now and then, however, a voice reminds us that over the long run nature can take care of itself. Brazil has moved to protect some 3.3 million acres of its Atlantic forest. The Serra do Mar complex of ranges, which runs for well over a thousand miles along the coast, has been mostly deforested, partly to make room for sugar and coffee plantations, partly because the lumber there is even more valuable than that in the Amazon, and partly by encroachment from Brazil's largest, most polluted industrialized area. The government will reforest some of that land, but much will be left alone. Paulo Nogueria Neto, Brazil's special secretary for the environment, plans to let the birds do the work there. "The thrush is the best reforester because he is efficient and free," Nogueria said. "Manmade reforestation is usually homogeneous, but the Atlantic forests are heterogeneous. If we want to repeat the same variety of plant life we should depend on the thrush," he explained.

About a week after I read that in the *New York Times*, I came across another scheme to let the birds do the work. This plan involves fertilizing rather than seeding. Coastal beaches and especially barrier islands are subject to erosion, both the steady lateral transport of sand and the explosive change wrought by storms. The dunes behind the beach are the land's first line of defense, ramparts against the sea. The dunes will march with the wind unless vegetation—sea oats and other salt-tolerant species—holds them together with roots. Dune grasses will seed themselves naturally. Humans also plant them, plugging clumps into bare dunes in the hope they will "take." The trick is to fertilize the plants so they form a thick cover, their roots forming a mat that will hold the sand against wind and storm. And that is where the birds come in.

John Hoke is a resources manager with the National Park Service, the same man who turned some of Washington's architectural pools into minihabitats where ducks now breed. Ten years ago he visited Pawleys Island in South Carolina, about two decades after Hurricane Hazel had torn most of the vegetation (not to mention most of the houses) away. He was impressed with how lushly the vegetation had come back, and began asking questions. The people there had re-planted grasses, it seems, later using conventional fertilizers to hasten regrowth. Even so, the growth struck Hoke as unusually lush, and he looked further. The difference between Pawleys Island and similar barrier islands, he surmised, comes from the number of birds—especially purple martins—there.

Pawleys Island is lined with purple martin houses, both the apartment houses so common farther north and arrays of gourds hung from yardarms on poles. The houses have been there for years, and are full season after season. Hoke found the most vigorous growth around the martin houses, both directly under them and within a circle stretching about 40 feet away from each house, which is about the distance parents carry fecal sacs produced by their young before dropping them.

It is Hoke's hypothesis that the martins provide enough continuing fertilization through their droppings to account for the lush vegetation. He has no numbers yet, but I have tried some back-of-the-envelope calculations, based purely on assumptions, to arrive at a ballpark figure. Suppose a martin house has 20 boxes or gourds. And suppose that 16 are occupied by martins (the other four having been usurped by sparrows). And suppose that each martin eats half its weight in insects each day, and excretes about half that. A martin weighs about 2 ounces, so each would produce half an ounce of droppings a day.

Thus our 32 martins are adding 1 pound of fertilizer a day to the area around their house. If they are there through April, May, and June, we are talking about 90 pounds. For July, double production with the addition of the young of the year, and we have another 60 pounds. Before this hypothetical colony leaves for South America in August or September, it may have contributed 150 pounds of rich fertilizer.

Hoke would like to try attracting martins to other barrier islands, especially places like the Outer Banks of North Carolina, where the government has decided to stop trying to arrest erosion by brute-force engineering efforts. He would like to see the residents there, aided and abetted by bird clubs and Boy Scouts, put up 500 or 1,000 martin houses and compare what happens with control areas in which there are no martin houses. It might take some time for martin colonies to establish themselves: The bird list for Pea Island National Wildlife Refuge on the Outer Banks lists the purple martin as an uncommon inhabitant in spring and summer (meaning "present, but not certain to be seen") and does not show it as nesting there at all. Martins are seen there mostly in the fall, during their migration to South America.

To me, having martins fertilize the dunes is as attractive an idea as having thrushes reseed the coastal forests of Brazil. Once you have put up the martin house, you can lie back in the hammock while the martins turn insects into fertilizer. The satisfaction would be like that I get from turning on the lawn sprinkler: No matter how much I goof off for the next couple of hours, I am getting something done. It is not yet clear that the martins would make any difference to the dunes. More has been claimed for them in the past than they actually deliver. The most famous claim is that a single martin eats 2,000 mosquitoes a day. This turns out to be highly unlikely. Herbert W. Kale II, a Florida Audubon Society ornithologist, looked into the question back in 1968. He found that published analyses of purple martin stomach contents revealed precious few mosquito remains.

More telling, Kale pointed out that the behavior patterns of martins and mosquitoes are so different that martins could not be expected to be major predators of mosquitoes. The birds are most active during the day, while the mosquitoes are most active at twilight and during the night. Their active times overlap only a little before sunrise and after sunset. Additionally, mosquitoes stay closer to the ground, rarely rising above the tree canopy, while martins spend a lot of time 100 to 200 feet above the ground. Finally, for martins to really put a dent in mosquito populations, Kale argues, they would have to be present in huge numbers (several thousand for each acre of mosquito-producing wetlands) when a hatch of mosquitoes first emerges. The claim is not completely false. Martins do eat some mosquitoes, and one road-killed bird was found to have 250 saltmarsh mosquitoes in its stomach. But they would appear not to be the mosquito scourges we have been led to believe.

Martins do eat insects, and a lot of them. A favorite food is dragon-flies. Another is bees. But martins also eat stink bugs, beetles, flies, spiders, grasshoppers, butterflies, moths, and bits of eggshells. The last is thought to be ingested because it provides a good source of calcium.

Mosquito killers or not, martins are nice to have around. So by all means, let's intervene. Not with bulldozers but with birdhouses. Will the purple martins slow the erosion of the barrier islands? Let's find out. Whatever the outcome, we have little to lose. While the birds convert insects into fertilizer, we will vicariously share in the excited hubbub of the martin house, where nonstop chirruping seems to be the continuous communication of breathless news.

Sitting outside after supper, watching the martins make their long glides across the yard to their houses, is not unpleasant. Sometimes they fly in circular patterns with their house at the center, a little like children dancing around the bandstand on the village green. The only drawback to having martins around the house, perhaps, is the sense of loss when the colony leaves long before summer is over. The yard seems unnaturally quiet, like a house when the children are gone. The martins are down in South America by now, giving the handy among us the time to build (and the rest of us time to buy) the apartment houses or the gourds and the poles we need. Next spring, when their scouts move northward with the wave of emerging insects, they will find—if John Hoke has his way— new places along the coast to nest and hunt and make the world green.

October 1985

Arbornauts in a Tropical Forest

 You are a bird, flying among the trees in extreme slow motion—silently, effortlessly moving in three dimensions. The law of gravity has been repealed as you pause in midair to watch ants around the sap oozing from a bud at the very tip of a branch, then soar up and over the tree to float down into a shaded cove on the other side. One moment you admire the iridescent colors on the tiny, stingless bees pollinating orchids; the next you hover above an iguana sunning itself in the tangle of vines that have taken over the treetops. A white-necked puffbird allows you to approach within 3 feet: It stares but does not move. A little farther away a black vulture pays no attention whatsoever. Then you go swooping off again, over some trees, around others, moving like a diver among coral heads. You drop down almost to the ground a hundred feet below and then rise magically, up and over the treetops. All too soon it ends. Your feet touch the dirt, and once again you are a creature of two dimensions rather than three.

Everyone has had the dream. A child looks up into the treetops and sees a world it cannot reach. An adult walks along a bridge that crosses a ravine and finds herself stopping where the ground has fallen away enough so that she is on a level with the treetops; her mind soars out into space. No birder can watch the objects of his affection for very long without realizing anew just how nailed to the ground we are by gravity, and dreaming for a moment of slipping the surly bonds.

Biologists have the same dreams, only more so. A tree is a device to capture solar energy: To really understand it, they have to be able to get to where photons meet chlorophyll. It is at the very top and out at the very tips that new leaves grow, flowers blossom, pollinators work their magic and herbivores wreak their havoc. Trying to work with only the flower petals, seedpods, and chewed leaves that fall to the ground, one botanist explains, is like being a marine biologist who works only with what washes up on the shore. For decades terrestrial ecologists have tried to rise to the challenge. They have climbed the trees themselves, risking close encounters with bees, wasps, and snakes. They have shot lines over high branches and hauled themselves up in bos'n's chairs. They have strung cables and even entire walkways from tree to tree. A group in French Guiana has experimented with what looks like a life raft that is dropped onto the treetops from a small blimp. All have limits to the access they provide.

Now biologists at the Smithsonian Tropical Research Institute (STRI) in Panama are trying a new approach. Allowed $40,000 in his budget to study the feasibility of using a tower crane for canopy research, Alan Smith, STRI assistant director for terrestrial research, simply rented one from a Panama City construction company and had it installed in a nearby park. The crane is just off a dirt road that follows the Las Cruces Trail, over which Spanish explorers carried gold and silver from Peru to ships waiting on the Atlantic side of the isthmus. For nearly a year now biologists of every persuasion have floated through the trees, placing their instruments and traps exactly where they have always wanted them.

The crane is just like those working in most cities around the world these days. This one stands 100 feet high, above much of the canopy of a dry forest that is only about 70 years old. The boom extends 115 feet, which means that the scientists in the gondola can move anywhere in a cylindrical space 230 feet across and 100 feet high. Jose Herrera, the operator, spent 4 years building high-rise condominiums in Panama City before he took the job in Metropolitan Nature Park. This job takes a lighter touch, he says: You can't swing a gondola of biologists around like a bucket of concrete. And unlike a construction site, which is clear of all obstacles save the building itself, Herrera often has to work without being able to see the gondola. Not only does he prefer the peace and quiet of the forest, however, he has even bought a pair of binoculars and keeps an eye on the wildlife while the scientists go about their business.

The gondola is a steel-floored wire cage large enough for two people and their equipment. The railing is about waist-high; above it are screened half doors on all four sides. Some of them are covered with aluminum foil as well to keep out the sun during a long session. A few scientists take a climbing rope with them for use as an emergency exit, but the only "emergency" in the first nine months of operation was a 30-minute power failure.

You get in the gondola on the ground, by walking up a few concrete blocks and then climbing over the railing. When everybody is ready it's thumbs up to Herrera, and suddenly you are rising through the air, in perfect silence, the understory dropping away beneath you. You glide upward among taller trees, seemingly unaffected by the large, pendulous ant and termite nests that hang off them. At the top comes the only surprise for a first-timer none too comfortable with heights: When the gondola begins to move out on the boom, it starts with a jerk that sets the gondola swaying—I clutched as tightly as I've ever held an airplane armrest. Otherwise you quickly forget the crane itself and stand facing the direction in which the gondola is moving, sailing over the trees as in a dream.

Now, finally, you are sunlight looking down on the trees. Here you are starting at the beginning and can hope for a new and better understanding of the most basic ecological cycles: the flows of energy and materials, the relationships of producers and consumers. To the eye of the totally uninformed observer, there is no rhyme or reason to any of it. No two species are doing things the same way. This is a place not just of competing trees but competing strategies. Some, for example, like that pseudobombax (a member of the kapok, or silk-cotton, family) over there, have dropped all their leaves in this, the dry season. Others, like the cecropia, with its hollow, chambered ant-condo branches, are just putting out new leaves on bare branch tips. Then there are trees putting out a whole new layer of leaves over the old, chewed ones, to capture as much sunlight as possible before the clouds of the rainy season roll in.

Even the leaves are totally different, from the feathery, mimosalike fans of the *corotu* to the heavy, elliptical leaves of the wild cashew (known as *espavé*, supposedly from the Spanish *es para ver*, "it is to see": The wild cashew was the tree Spanish explorers climbed to see where they were). In the tops of some trees, the woody vines known as lianas have grown so thick that the underlying tree would appear to be in mortal danger. And this was one of the first surprises to come from going up in the crane, according to Alan Smith. More than 35 percent of the canopy is occupied by lianas, he said, far more than expected. It had been the conventional wisdom that because the crowns of trees sway in the wind and abrade their neighbors, there would have been selection for leaf toughness, so one tree could grind out more space for itself. But the lianas lace the treetops together so that several trees move as one and there is little abrasion.

Another surprise was the extent of herbivory. When an entire leaf is eaten at the top of the canopy, a biologist on the ground has no way of knowing it ever existed. If STRI can install a crane permanently, Smith said, biologists will be able to measure the herbivory and identify most of the perpetrators. One perp leaves a trademark: Leafcutter ants cut a scalloped pattern along the central rib of the leaf; they laboriously carry the small pieces of foliage all the way down the tree and across the ground for many yards to their nests.

Basically, the plant ecophysiologists want to know more about how tropical plants respond to their environment: Why, for example, do some trees in this always warm region drop their leaves during the dry season? It could be a way for the tree to rid itself of all its herbivores and their eggs in one fell swoop. Or it might be a way to stop cavitation, a life-slowing situation in which the tree's water-circulation system breaks down. Trees pull water up from the roots through a system of vessels called xylem. During the dry season, however, the pull may break these thin columns of water inside the tree, leaving embolisms. Dropping leaves stops the pull and avoids the danger.

Other questions of great interest are how plants cope with the potentially damaging effects of strong light and low moisture (in the dry season) on the uppermost leaves of the canopy. Some plants simply turn their leaves so as to absorb less light. Others are able to shunt excess energy from critical leaf enzymes. Strong sunlight is beneficial, of course, if the leaves can stand it and can make enough enzymes. It is costly, however, for plants to provide the extra enzymes needed to take advantage of such high light levels, and they may lose much of what they accomplished in daylight during nighttime respiration.

Each day that I was in the gondola, two Panamanian research assistants from STRI worked on the ground beneath us, measuring photosynthetic activity of leaves in the understory. Their instrument encloses a leaf and subjects it to different levels of light, carbon dioxide, and humidity. It then measures the resultant activity. When we left, the women would rise into the sky to make the same measurements in the canopy.

On my final flight with the arbornauts, ecologist Kevin Hogan talked about finding out what attacks flowers and seeds at what stage, and especially about predation on seedpods: beetles that bore their way in, monkeys that tear them apart to get at the grubs laid by the beetles. Some trees tend to build very hard, protective pods and only mature the seeds at the last moment, he said. But no one can yet say whether a tree drops an insect-ridden pod because it is insect-ridden.

The crane has already been used to map the forest canopy. Geoffrey Parker, from the Smithsonian Environmental Research Center in Maryland, and Alan Smith moved the gondola along the boom, recorded the height and species of the canopy below, then moved the gondola a few feet and repeated the process. At the end they produced a "topographical" map of lianas and trees.

That afternoon I boarded an old excursion boat in Gamboa, rode up the Panama Canal for an hour, and stepped ashore on the hallowed ground of Barro Colorado Island. On any map of where tropical biology is done in the world, this is one of the major capitals. A research station since 1923, its trails and buildings are named for the giants of the field who worked here, now to be met only in textbooks. And it is here in this rain forest, so well studied from the ground, that STRI would like to build its own permanent tower crane.

The people here have tried other ways to study the canopy. George Angehr, an expert on hummingbirds, led me up a 138-foot meteorological tower, a steady staircase inside steel scaffolding. (He tells the story of watching a hummingbird leave a nest 70 feet up and simply drop toward the ground, fluttering from side to side like a falling leaf. Near the ground the bird suddenly shot away in normal flight. The behavior, presumably, avoids calling attention to the nest.) On the way up I saw that almost every leaf within arm's reach had been tagged and numbered, but at the top the foliage is out of reach; all there is to do is hang on tight and enjoy the view. Another stairway to the sky is even taller. Gerhard Zotz, a German researcher, has lashed a triangular radio mast to the side of a ceiba tree. It's about a 110-foot climb to his platform on the lower branches (he is studying a hemi-epiphyte, an air plant that dropped roots to the ground and has become a tree) and another 45 feet up the tree to a vantage point above the canopy. (Even with his kind offer of a climbing harness, I made it very, very clear that my tight schedule did not permit a climb that day.) In the 1920s, another researcher, I was told, had simply hammered railroad spikes into the side of a tree and climbed them like a ladder. But none of these towers moves.

Walking the trails of Barro Colorado, I looked in wonder at keel-billed toucans and white-faced monkeys, delicate air plants and the 20-foot-high buttresses of ceiba trees. For someone who has spent decades editing other people's stories about tropical forests. it was a dream come true. Yet as I plodded along, sweating happily, I found myself peering through the trees for the bright orange, totally unnatural crisscross pattern of a tower crane. How're you going to keep us down on the ground, once we've been up in the trees?

June 1991

Prairie Grass Where Protons Fly

 Once this place was tallgrass prairie mixed with oak savanna. Within a few years it will be again— one of the very few such places in Illinois, in all of the Midwest. Humans have been busy here for the past 8,500 years, but not until 1830 was the prairie plowed, the trees cut. Now a small army of scientists and engineers is restoring what once was there, burning and seeding the land.

For the past 15 years it has been the site of one of the great achievements of the human mind. In the middle of the 6,800-acre tract is a circular berm, a ridge 4 miles around. Beneath the berm is a tunnel, and inside the tunnel are two metal tubes, one over the other. In the lower tube bunches of protons and antiprotons, each as long as a pencil and as thin as a human hair, plow into each other at nearly the speed of light, releasing energy in trillion-electron-volt bursts. Physicists are using what today is the highest energy accelerator in the world to look for the "top" or "truth" quark, the last of the six quarks they believe are the basic building blocks of matter.

This is Fermilab, the National Accelerator Laboratory, 27 miles west of Chicago in what once was the farming community of Weston. I was there to visit the natural habitat of Jim Trefil, a theoretical physicist and writer. He had once joined me in my preferred habitat, the Atlantic coast at Chincoteague, Virginia. Now I was on his turf.

Our guide was J. Richie Orr, the associate director for administration, just back from Washington where he and three colleagues had received the National Medal of Technology. Orr led the team that devised—with their hands as well as their heads—the innovation that doubled the energy levels the Fermi accelerator was originally designed for. Orr's team figured out how to pull the wire and wrap the coils for 1,000 superconducting magnets that make up the new ring through the tunnel.

Accelerators work because subatomic particles with an electric charge, positive or negative, respond to magnetic fields. As the particles race around the ring, more and more electricity is pulsed through the magnets, kicking the particles harder and harder. The faster the particles go, the more energy it takes to control them, to keep them in a usable beam. Any moving object tends to go in a straight line; that is why the passenger is pushed against the right-hand door in a car turning sharply to the left. In a circular accelerator like Fermilab, the particles are constantly being turned; it takes huge amounts of energy to bend the beam. The lab was paying $16 million a year just for electricity.

In the original ring at Fermilab, a thousand conventional magnets keep the beam in place. Energies of 400 billion electron volts are achieved. But in superconducting materials electrical resistance drops to zero. Magnets made of such materials can produce and control much stronger beams for a fraction of the electricity. The new ring produces energies in the trillion-electron-volt range.

Orr led us along the same path an unsuspecting proton would follow. The process begins inside a Cockcroft–Walton generator, where electrons are added to electrically neutral hydrogen atoms, giving them a negative charge. They come out at the speed of sound. Next they enter a 500-foot-long linear accelerator, where radiofrequency waves oscillating through 290 electrodes further accelerate the particles and strip them of their electrons, leaving only positively charged protons moving at a substantial fraction of the speed of light. The rig is powered by a 5-million-watt amplifier (a device my rock-musician son would kill for).

The protons next find themselves in a circular accelerator, where still more energy is added, and then they are injected into the 4-mile tunnel, in the upper, original, ring. They are accelerated still further, then dropped down into the new, superconducting ring. At first the protons are in bunches the length and width of a pencil. Each bunch contains about 10 billion protons, which sounds like a lot but is really almost empty; normal matter would have more like 100 trillion billion protons in the same space.

Some of the beam is diverted into targets that produce antimatter, specifically antiprotons. These are accumulated in a small storage ring; after about 4 hours, enough have been collected for the real business to begin. The bunches of protons in the superconducting ring are pulled tight so they are the thickness of a human hair. Bunches of antiprotons are injected into the upper ring, accelerated, then dropped into the lower ring, traveling in the opposite direction. There are three bunches of each, separated by 120 degrees. The two sets are now accelerated to a trillion electron volts and steered into each other.

For the physicists sitting in a cramped control room, watching the action on 60 different screens, the trick is not just to have the beams collide but to have the collisions occur in the exact center of huge detectors. They occur 50,000 times a second, and it takes three floors of computers to sort out what is happening. The collisions create, for a very brief time and in a very small space, conditions that existed early in the formation of the Universe, as early as one ten-trillionth of a second after the Big Bang. "We are seeing matter in a form we have never been able to see before," Orr explained. "We create a hot, early stage, and then watch it cool."

On top of the ground, high-energy biologists are also engaging in time travel. They are restoring the tallgrass prairie that once covered this land from horizon to horizon. Since the Nature Conservancy helped them initiate the project in 1975, they have been collecting seeds from nearly extinct plants that once grew in such profusion, finding them in undisturbed cemeteries and along railroad rights-of-way, and planting them inside the ring.

Slowly, like Brigadoon appearing from the Scottish mist, a prairie is taking form. Robert Betz of Northeastern Illinois University, the guiding force of the project, led us through the bluestem grass and the compass plants, the indigo and the prairie dock. His co-workers have restored 700 acres so far, he said, and are still adding two or three species a year to the mix. It takes at least 3 years for a new section to build up enough fuel for the fires essential to restoring a true prairie.

They hope to reach at least 2,200 acres, although Betz concedes it would probably take many more to approach a true ecosystem. Some organisms aren't waiting: Seven pairs of red-tailed hawks now nest at Fermilab.

The prairie rather than physics may be the future of the site. As part of his tour, Orr led us through the shops where the superconducting magnets were built and are repaired. In one room we were confronted with a 57-foot-long red cylinder, looking something like a huge torpedo, that had the letters "SSC" painted on the side. It was a prototype magnet for the Superconducting Super Collider now being planned for a site in Texas. The SSC is the next generation in accelerators, so big that a ring the size of Fermilab's will be used as an injector.

Orr has mixed feelings about his shops being used to build magnets for the machine that will put his out of business, about the Fermilab beam being used to test detectors for Texas. He talks, not quite seriously, about retiring, perhaps to teach. But the succession of larger and larger machines is the nature of the business.

In its present configuration, Orr estimates, the Fermilab accelerator can do useful research for another 5 years. If the lab gets the money for a new injector ring, so that instead of having three bunches of protons collide with three bunches of antiprotons there can be 32 hitting 32, then there will be 10 more years of productive work. But eventually the accelerator will be history, and the exploring will be done in Texas or elsewhere.

Future visitors will see a time travel machine and travel in time themselves. They will tour the accelerator that enables scientists to see matter as it was in the early moments of the universe. And when they walk the prairie above, they will be seeing Illinois as it was before Europeans arrived with plow and axe. Accelerator and prairie are both tributes to what is still the most impressive "machine" we know: The human mind.

June 1990

The Wild Is All Around Us

 My friend laughs a little when I head out the door to get back to nature. She has not been the same since I took her to see one of the 200 saltwater crocodiles still hanging on in Florida. A local expert on Key Largo had agreed to take us; I assumed we were headed for the new national wildlife refuge established especially for these endangered throwbacks. Instead, after winding around back roads, we arrived at a run-down marina. Our guide looked crestfallen. "Usually there's one sunning itself on the boat-launching ramp," she said.

My smiling friend knows that my idea of getting back to nature has little to do with what John Muir used to do. The places I go have trails well marked with both signs and beer cans. Boardwalks appear where the ground is damp. A nature center on the grounds has proper restrooms. I feel a little silly myself. We are both victims of an ideal: that wildlife should be seen in its own Edens, places where humans have had no impact. An unnamed lake in the Canadian north, perhaps, or a little-known valley in the Andes.

The wild creatures themselves are not so fastidious. More of them than we might expect have learned to live with us. I do not mean the rats, roaches, and sparrows that live with us everywhere or even the spiders, mites, whiteflies, and aphids that find homes on our houseplants. I mean those wild creatures.

A case in point. In my part of Washington old trees line the streets; each house boasts a postage-stamp yard in front and a slightly larger yard in back. Lots of us feed the birds through the winter. Early last spring I noticed something alarming. In the morning, when I walked the dog around the block, the noise was phenomenal. Birds were seeking mates and staking out territories so vociferously that I began to wonder about ear protectors. Some days I would take the dog six blocks down the hill to where an abandoned mansion abuts a park. This was real habitat: some towering old trees, lots of new growth and understory, overgrown fields, a robust little stream. The dog and I would slip through the iron fence and walk down to the stream . . . in perfect silence. Not a note to be heard, not a bird to be seen. Every last one was up among the houses, the people, the cars gunning to make the light.

Curious. As spring softened the world, I heard of mother ducks leading their new broods across six-lane boulevards downtown, making their way from little artificial ponds in front of government buildings to the muddy embrace of the Potomac. Beavers invaded backyards in the Virginia suburbs, dropping prize-specimen trees on manicured grass. Interesting. Only when I sneaked in a little birding on a couple of business trips did a pattern suggest itself.

In Florida I rushed from Miami down Route 1 toward the Keys, scanning the telephone wires and ditches along the road as I drove through a remnant of the Everglades. White wings through the reeds caught my eye at one point, and I stopped to check out the egrets feeding in a ditch. There, in all its pink splendor, was a roseate spoonbill, huge next to the egrets, swinging its spatulate bill from side to side in the muddy water. I was not to see another in all the protected parks and refuges from there to Big Pine Key.

Another day I went north, to the Loxahatchee refuge. I dutifully walked the dikes around the impoundments, happily watching the ducks and blacknecked stilts. Back at the headquarters, next to the bustle of the parking lot, I looked up to see large white birds with naked heads in the tops of the trees: wood storks, an endangered species. Eleven of them perched there, while another three flew in circles.

Then there was a week-long meeting in San Diego, in a hotel on filled wetlands in Mission Bay. The hotel had gone to extraordinary horticultural lengths, planting tropical trees from around the world. In the middle was an elaborate pool in which the water was a porcelain blue, the result of a dye used to discourage algae. It looked totally artificial and in my wisdom I decided to have nothing to do with the ducks and coots that would paddle frantically toward anyone who paused by the 25-cent feed-pellet dispenser. My moral superiority meant that for 2 days I failed to notice the herd of black-crowned night herons that spent their days on an island in the pool. Finally I looked, and saw also a couple of dozen lesser scaup (a sea duck), the wildly painted eye of an American widgeon, a wood duck couple just coming into their nuptial splendor, a scattering of Mexican black ducks, and a double-crested cormorant that appeared to be fishing successfully in that strange blue water.

One free afternoon during the conference I rented a car and headed south. I tried the Tijuana River National Estuarine Sanctuary, where the river comes out of Mexico to empty into the Pacific. The river itself gives you the impression that you would not want to swim in it. Worse, you cannot hear yourself. Abutting the refuge is a naval air station where helicopter pilots train: All day long the huge machines beat out over the beach and back. A horrible place to go bird- ing . . . except for the birds. The muddy banks of the tidal channels were loaded with willets, whimbrels, godwits, and a still more exotic bird I had never seen before: the long-billed curlew, the bearer of a down-curving bill 8 to 10 inches long.

By then it was dark and windy, but I rushed to one more place, the southern end of San Diego Bay. Shivering, I headed out onto the dikes around a salt evaporation plant. The pickings were poor: dark ducks at a distance, an occasional cormorant diving. I turned with the dike and found myself headed for shore at the very bottom of the bay, walking above some sort of drainage channel. The tide had run out, exposing dark mud and the usual trash. And then, there just ahead of me, delicately foraging, were half a dozen birds of a species I had been hoping for years to see: the American avocet, a good-size wader with a long bill that curves up near the end, the head and body a striking black-and-white pattern with the pinkish tan of breeding plumage taking shape. In an ugly place, a beautiful creature carried on as it had for thousands of years before humans assaulted its home.

Late on the last afternoon, I wandered over to the marina in the next cove. Out in the channel floated a long row of bait boxes, their tops just above the water. And standing shoulder-to-shoulder along the entire row was the largest collection of great blue herons, great egrets, and snowy egrets I have ever seen. In front of me the crew was cleaning up a charter fishing boat. A brown pelican waited in the water for scraps. But the crew kept pushing the pelican away with a broom: They were feeding the scraps to a sea lion. It reminded me of the Keys, where normally shy great blue herons invade backyards and motel grounds looking for handouts.

Life pushes against the boundaries in every direction, including straight into the strongholds of civilization. Birders drive out of cities to find the unusual, the spectacular, the rare. But the birds themselves push into cities, looking for Lebensraum. Loren Eiseley once wrote of birds perched on a television antenna, asking each other: "Are they gone yet?"

The question that comes to mind is whether we have so impoverished the natural world that more and more wild creatures are learning that they must approach humans to find what they need. (Those panhandling herons in Florida raise young more successfully than their counterparts in the wild.) Perhaps our cities are becoming to wildlife in general what the Yellowstone garbage dumps became for bears. Whatever is happening, it behooves us city dwellers to keep our eyes open.

July 1988

PART 4

The Ecology of Human Beings

 A single human being is an exercise in ecology, trillions of cells all connnected by what is now coming to be known as the psychoneuroimmunological system. The adjective compounds what had always been thought of as three separate systems that did not interact with each other; now it seems they may be actually different manifestations of one, grand system. (This leap in understanding is comparable to the physicists' showing that the weak and the electromagnetic—and almost certainly the strong—forces are actually one and the same. They are having less luck with the fourth fundamental force, gravity.) Just as we are coming to understand what we are doing to nature and what we can do about it, so we are learning more about ourselves.

By divine coincidence, the ecologies of the natural world and the human body overlap. Medical authorities are now conceding that you don't have to get your heart rate up to 80 percent of maximum for exercise to have any benefit. Plain old walking is almost as good. So every time you walk instead of drive some place, you do something for both ecological systems. (Bicycling multiples the range of virtue.) When you have the pasta primavera instead of the porterhouse, you do something for your arteries and for the land.

Ecology, like charity, begins at home. How are we going to take care of a planet if we do not take care of ourselves? A heart attack is like a fish kill: a signal that something has gone radically wrong. It is fun to learn that snacking and napping may actually be good for us; it can be life or death to learn how to handle stress, to gain control of our time and our lives.

Taking better care of ourselves is even patriotic. One of the reasons for the horrendous medical costs in this country is the enormous percentage of medical problems that are self-induced. We are eating, sitting, and smoking our way into the hospitals, where billions are spent to undo the damage we have done. There must be a better way to spend that money.

This is not a call to asceticism, however; I may be the least qualified person in history to issue such a call. Rather, it is a suggestion that it can be fun to try to figure out all the ways we have an impact on the Earth, good and bad, and on our own well-being, and then try to see the connections. And the more we appreciate both the natural world and ourselves and other human beings, the more the game of figuring out how we can best serve the two ecologies can be a labor of love.

Voting with Our Feet

"It would kill any of the young men of the present day to attempt such a walk; it must be four miles, or two, or some immense distance."

Those lines, spoken by a noblewoman in Emily Eden's 1830 novel, *The Semi-attached Couple*, somehow strike me as funny in this day of 3-hour flights to France—funny at least until the next time I find myself in a 4,000-pound car carrying four pounds of milk four blocks. On the days when I drive to work (the days when there is not time for the subway: Mass transit it may be, but rapid transit it is not), my mind is too busy reviewing the roll call of anxieties to take heed of any other driver not actively trying to run me off the road.

And then my father will be up for a visit from South America, I'll take him downtown in the morning, and invariably he is stunned at the sight of all those cars—almost every one carrying one human being and no more. He lives in a place where people live environmentally sound lives for the wrong reason: They are too poor to do anything else. When he has an empty jar or bottle, he simply sets it outside the garden gate. Within five minutes it has been recycled, pressed into immediate service.

Actually, I was a little stunned myself when I first started working in Washington. I would come out of my building at the end of the day to find the sidewalks almost deserted. Our magazine offices are in a building on the Mall, which of course is a special case: aside from the concession stands, this is a noncommercial district. That means that it is a desert as far as human amenities are concerned. Nowhere can you get your glasses fixed, pick up some flowers, or buy a bottle of aspirin.

Now the Smithsonian itself employs only a few thousand people.

But major federal buildings line the avenues to the north and south of the Mall. Altogether we are talking about a fair number of people. Yet when you come out of the back of our building at 5:30 or 6, and look across the street to where all those people have been spending our tax money all day, there is hardly a soul in sight. The sidewalks are almost empty. Everyone disappears underground to catch the subway or to get into a car and drive away. Even those who want to get together to discuss the latest managerial madness get in their cars and go somewhere else to do it. After working in New York, where the sidewalks are alive, this absence of people produces a weird feeling. You begin to wonder if you have stumbled into one of those movies in which something terrible has happened to everyone except the hero.

The good news is that we can turn places like Washington into real cities and at the same time ease two national problems without spending a penny. As a nation we worry about global warming; we fear the ozone on the ground that injures our trees and our lungs and the thinning ozone miles over our head that no longer protects so well against ultraviolet radiation. We worry about the nitrogen and sulfur oxides that fall again as acid precipitation and the carbon monoxide that crowds the oxygen out of our red blood cells. We're doing it to ourselves, every time we put the key in the ignition and turn it.

As individuals we also worry about our health and the urgent advice of the authorities to do grim penance for the french fries of long ago. We nod our heads and tell friends we really should get out and jog, join that health club, buy a $600 machine for the bedroom. But we are rushed this morning, tired tonight. Next week for sure: We're going to get right on it. And sometimes we do, and then drop out.

Let's leave weakness of character out of this. People drop out because that kind of exercise is boring, makes you ache and, if you don't happen to be a gazelle anymore, rubs your nose in your own deterioration. It has no purpose beyond the motion itself. On one of my very few excursions on horseback (Cocoa, wherever you are, thanks for keeping us right side up!), I liked it when the wrangler stopped snickering long enough to explain that while Eastern riders do it for show, out there in the West people ride horses to get somewhere.

I've always thought it made more sense to use gardening and yard work for exercise rather than pay to go to a club and pull weights on wires that run through pulleys. When you finish your garden "workout," you have accomplished something. News reports years ago of Chinese office workers being forced outside to shovel snow struck me as an inspired idea. What an incredible antidote for the post-power-lunch mid-afternoon slump.

Now what I'm driving at here (pardon the verb) is a way to exercise that does not require adding a whole new, unpleasant chapter to each day. Something that can be done without special equipment or going to a special place. Something that lessens the damage being done to the planet we love. It seems that you don't have to get your pulse up to some very high rate and keep it there for so many minutes so many times a week for exercise to do any good. Any exercise is good for you, we now learn; one of the best is just plain walking. All we really have to do in cities is what the National Park Service is forever urging us to do when we are out in the middle of nowhere: get out of the car.

The kind of walking I'm talking about has nothing to do with hardship and self-discipline. I'm talking about city walking, a delight for body and mind. Just suppose you found yourself walking 5 miles a day. Not all at once, mind you, but a little in the morning, a walk at lunch, a trip to the post office; later a walk part of the way home, even down to a restaurant or the movies in the evening and then home again. Find yourself noticing cloud formations or swifts spiraling down into the library chimney, bats flashing under a streetlight. Check the bookstores, the notices stapled to telephone poles. Find that faces are considerably more interesting than bumper stickers.

Now I am talking about plain old walking here, jaunts of up to, say, 5 miles. Hiking is something else. A person standing in Maine about to head down the Appalachian Trail to Georgia needs special equipment, training, experience. That's a whole other world. I don't know what it means to walk 20 miles in a day, and certainly not with a pack on my back. I admire those who can from the same distance as I admire carrier pilots and brain surgeons. When I finally read Edward Abbey's *The Monkey Wrench Gang*, I was as taken by George Hayduke's ability to throw a 60-pound pack on his back and walk the otherworldly terrain of Arizona and Utah as I was with the gang's adventures. (Later I read of John Muir's taking off into the mountain wilderness for three-day junkets with just a little tea and bread in his pockets, and thought there must be never-ending levels of self-sufficiency.)

My kind of walking is restful. No one is climbing up your back, leaning on a horn designed to blast an opening a mile ahead while doing 130 on the Autobahn. If something catches your eye, you can stop without being rear-ended. You discover that you do not have to be training for the Iron Man Triathlon to release a few endorphins in your brain, letting the sun come out in your mind. How many times have I started out to walk somewhere, feeling tired and irritable, begun to think the whole idea was a dumb mistake, then found that despite myself the slump was coming out of my shoulders and I was deciding not to have that troublesome writer done away with after all?

Governments dither, but there is nothing to keep us as individuals from voting with our feet. We can do something about global warming, and feel better for doing it. Each time one of us walks someplace instead of driving, that much less poison has been injected into the air we all breathe. (Actually the benefits are multiple. One less car means that the rest of the traffic will move that much more quickly and burn less fuel.) Now suppose two of us walk someplace instead of driving. What if 100 million of us did it just once a day?

As time went by, a third benefit would accrue. Paradise is not now upon us. The streets of Washington, like those of most other cities, are not yet a permanent festival, promenaded by philosopher kings. In Washington, the tide is still going out. The corner restaurant where you could sit in a booth for hours, the bakery where people lined up for the strawberry pies, the health food store that was forbidden by law to use the word "health" in its name but sold a ton of the No. 7 salad (chopped carrots, raisins, and cottage cheese, with a mayonnaise-and-buttermilk dressing) are all disappearing as the old buildings come down and great glass blocks go up in their place. As urban sociologist William H. Whyte says, "prosperity [is] lowering our real standard of living." But if hordes of us take to the sidewalks, we can return these mean streets to their highest function, the bringing together of human beings.

When we reclaim them, the sight of armies of potential consumers on the march will not be lost on entrepreneurs. They will become placer miners, dipping into the stream for gold. After a couple of quick trips to Rome, what I remember most fondly is not the churches, the shops, or the Colosseum— but the bars. These are not establishments devoted to the consumption of alcohol but rather rest stops, places where on the way to work, or home, or your next appointment, you can stop in for a small coffee, a delicate pastry or a sandwich fresh made with whatever is in season. Places where you and a friend can go to talk. There seemed to be one in every block, and it is like having a butler who is ready at any time to produce a little something to tide you over. No one would dream of driving to one; even if you survived Roman traffic, there would be no place to park. You walk there. The day may come when we can walk to one here.

Look, I'd really like to finish this story, but I've got to get up to 19th and E for a lunch date. I guess I'll walk on over to the Museum of Modern Art of Latin America, then go by Bolivar Pond to check on the ducks. After lunch I'll hike up 17th to pick up tickets for the ball game, and then come back down 15th, past the Boy Scout statue and on back home to the Mall.

July 1989

The Perils of Making Tea

 Back in the bad old days, before the battle of the sexes had come to such a harmonious conclusion, certain taunts exemplified pride in gender-based abilities. Female persons were wont to tell male persons that their uselessness in the kitchen was near total: "You can't even boil water." Ridiculous, of course. Anyone who understands throw weights, stellar collapse, and wild-card playoff berths can almost certainly boil water. The real issue runs deeper. There is more to boiling water than pouring some in a pot, putting it on the stove, turning on the heat, and waiting for bubbles to appear mysteriously in the depths, careen dizzily to the surface, and there silently explode into the steam that makes a kitchen alive. Female persons used to do that all the time, without a second thought. But it is the thought that counts. The decision to boil water raises questions that go to the core of style, of human expression, of who and what we really are.

Let's say we have been out shopping; we're tired, and a cup of tea seems just the thing. It is a simple matter to walk into the kitchen, step to the sink, turn on the tap. There is a high probability, as a scientist might say, that water will emerge. Just plain water. No French bubbles. No "natural" lime flavor. Just water. Merely a simple combination of the primordial stuff of the Universe (hydrogen) and a hyperactive refinement (oxygen) formed deep within a star and then flung out into space for the benefit of the next generation. What do the sociobiologists have to say about stellar altruism? (Waiter: "Would anyone care for something from the bar?" Patron: "Not for me, thanks. Just bring me something from an exploded star.")

Some of the liquid coming out of the tap, as befits its cosmic origins, is very old—water that first reached the surface of the Earth eons ago in a rain of comets or the outgassing of primordial volcanoes. Since then it might have spent a million years here, a million there: percolating through the ground until it gushed forth in a spring; moving with a cold current, deep in the ocean. More recently it has evaporated from the surfaces of oceans only to fall as rain over land, been breathed out by a tree and then rained down again to be drunk by a swallow on the wing. It has been vapor in the tropics and a rigid lattice in the Arctic. Now it has fallen as rain in the watershed of a river and been diverted into the apparatus of a municipal waterworks.

But suppose we stop to think about what else will come out of the tap. Will we get pure, life-sustaining liquid—or a thin feedstock for a petrochemical plant? Undeniably, there is water coming out of that tap. And some air dissolved in the water. And the dissolved minerals that give it flavor. (Do you prefer your water, like your scrambled eggs, hard or soft?) To make sure nothing unpleasant will happen to us when we drink it, that water also includes the chlorine that killed the germs. (Pour a little directly into the aquarium and the fish will die. But not to worry.)

So far, so good. We could be in grandmother's kitchen. This is a more complex age, however, and the list of what could be in the water is long. Nitrates, pesticides, and herbicides from farm fields. Nitric and sulfuric acids, raining down from the sky or trickling out of the hills where coal was stripmined. The industrial compounds put to work since World War II: Polychlorinated biphenyls (PCBs) come to mind. So do dioxins: These Cadillacs of toxins are found in the atmosphere everywhere around the globe, so it seems reasonable to expect they are in the water, too. Where did I read that many water departments still test only for those pollutants known and feared in the 1890s?

Well, look, it's only a cup of tea. Nothing to get all worked up about. Let's plunge ahead and actually run some water into a pot. A quart, say. That does not seem much at all—until we think about it. A quart of water weighs roughly two pounds or, in the units that are used all over the world (except, of course, in Burma, Liberia, and the United States), about 900 grams. This is about 50 times the weight of one mole of water. ("Mole" is a chemist's term for a large quantity of atoms or molecules; it is a useful way to deal with large numbers, as is "light year" for astronomers.) Multiplying the number of molecules in one mole of water (Avogadro's number) by 50, we come up with 30,114,800,000,000,000,000,000,000 molecules in our pot, all held together in the delicate embrace of hydrogen bonds. We are going to use heat to agitate these molecules violently enough to break uncounted millions of those bonds. If T. S. Eliot had J. Alfred Prufrock hesitate to eat a peach, shall we lightly tear asunder this many joinings, formed by laws of nature old before the Earth was born? Should we not at least think about it first?

We can't stand here in the kitchen all night. If we are going to do it, we need something to do it in. Shall we use this everyday pot—and run the risk of aluminum atoms clustering like grapes in our neuronal junctions, possibly leaving us with the horror of Alzheimer's? What, then? Good German iron? Glazed French clay, leaching lethal metals into the tea water? The sterility of glass? Man does not live by water alone, of course. What are the esthetics here? With the material chosen, or settled for, what of the design of our pot? The shape of tomorrow from a fevered mind in Milan? Or perhaps a pot of the ages, a form familiar in peasant cottages for a thousand years? As the Count de Buffon might have said, "The style is the male person himself."

Tea is a contemplative drink, and there are the questions of society at large and the future of our species to consider. Do we have the unbridled audacity to heat on an electric range? If those generators are turned with the steam from nuclear fires (now *there* are people who *really* boil some water), should we mail a check for our share in the cost of the future decommissioning of that plant, covering it in concrete and guarding it for 25,000 years from our curious descendants, if any? If by coal or oil, do we have a clear picture of exactly whose air and water we will poison to brew our tea? Or do we have the laboratory efficiency of natural gas, the bright blue flame of life long past? Shall we return to our lost Eden, using wood the way our people have always done? Sustain ourselves with broken bits of the living Gaia? We must always be wary of culture clash: Is it politically correct to heat a red-enameled Italian pot over mesquite charcoal?

It is nearly time now. Ignore the impatient calls from the living room (what can she mean, "dithering"?). We must resolve the question of speed. Do we go for the quick, unexamined boil or the teasing, sensuous rise to the magic temperature that prolongs pleasure, satisfies some unarticulated yearning in the soul? Great chefs will argue about whether meat should be cooked fast or slow; should not the same question arise in regard to water, far more fundamental to our very existence? Do we know how long it takes to boil the deadly chlorine out of city water? Or how long we can heat noncarbonated bottled water before we have driven off all the dissolved air and made it flat? Must we, as always, compromise? Do we join the self-torture of the high school physicist about whether cold water will come to a boil faster than warm?

It all seems too much for tired shoppers. Paralyzed, we stand in the kitchen, chilled, peckish, the weight of the world on our shoulders. China gleams, tea in its canister offers equatorial sun. The promise of soothing warmth. A hand reaches for the kettle . . . and stops in freeze-frame. The taunts of female persons come true. We cannot do it. It is not a question of somehow using up a limited resource. Just as would happen on any other well-run spaceship, the water we chose to boil would be recycled. Some would escape as steam to humidify the air in a winter house, to be breathed in and out until finally it is exhaled outside and escapes back to the sky to rain again on some other continent. The part we actually drink would also be recycled, although at somewhat higher expense to society. It, too, would eventually return to the ocean, to the sky, to another river so that someone else— next week or next millennium— could pour it into a pot and brew up some tea.

No, we cannot do it because too much is at stake here. In these troubled times, when death lurks in the very air we breathe and in the water we drink, when the fate of the planet hangs on our smallest actions, to boil a quart of water for tea would be going too far.

December 1988

Time: The Nonrenewable Resource

Every January 1, like the ne'er-do-well children of a wealthy family, we are presented with a handsome sum (more than half a million) which is ours to spend during the next 12 months. Each of us is given, not the dollars we fantasize about, but 525,600 minutes with which to build a life. We do think about it like money sometimes—we speak of spending more time doing this or that. But much of the time we don't think about time at all.

Time is like money in other ways. At least for those of us of a certain age, inflation is working its dark magic: Minutes (or hours, or days) are not what they once were. They lose their value. Time goes by faster and faster, days blur, season tumbles after season. In the tidy microworld of physics textbooks, time slows down for objects moving at relativistic velocities. In our messier macroworld, Einstein's equations do not hold: The faster we go, the faster time goes.

We can compensate by trying to do things faster, by cramming more into the same space. Some do this extraordinarily well. A defensive back in the National Football League can change direction (at top speed) three or four times in the second or two it would take me to (1) realize that the person coming at me has turned to his left and (2) send electrical impulses down my potholed neurons telling my legs to push me off to the right. Just watching them warm up before a game, backing up in quick little side-to-side jumps, makes a person realize how ponderous the rest of us really are.

Animals get even more out of a second. Consider a dragonfly darting, stopping, changing direction, as it hunts over a pond. One of its seconds must be longer than one of ours. Experiments leave no doubt that a second must seem longer for a pigeon, for example, than for a person. One researcher taught a group of pigeons that a flashing light meant food was available, while a steady light meant no food. He would then increase the tempo of the flashing to see how fast it had to go before the birds could no longer distinguish flashes. Long after the human eye would see a steady light, it turned out, pigeons are still able to see the flashing.

We try to take a page from those defensive backs and train ourselves to move more quickly, to be able to do more in a minute or an hour. We walk faster, talk faster; in our cars many otherwise calm people become Type A personalities, furiously impatient at the slightest delay. This kind of hurriedness is not only an intuitive perception; as we watch our fellow citizens surge along sidewalks, it can be measured. Robert V. Levine, a professor of psychology at California State University at Fresno, has done that, comparing America with other cultures and also comparing different regions of the United States. He summarized the results in *American Scientist*. To compare countries, he and his associates measured how fast pedestrians walk on downtown streets, how long it takes postal clerks to fill a request, and how accurate bank clocks are. Of the six countries he studied, the pace of public life was fastest in Japan, followed by the United States, England, Taiwan, Italy and Indonesia. In each country, the pace in larger cities was faster than that in smaller ones; this difference was most pronounced in the least-developed countries.

In 36 cities in the United States, Levine and his colleagues used four measurements: walking speed, talking speed, the time taken for a simple bank transaction, and the percentage of people wearing a wristwatch (a measure of their time consciousness). Boston, rather than New York, turned out to be the fastest-paced city of those studied; Los Angeles was the slowest. Chicago falls in the middle, while Salt Lake City is right behind New York. The Why is more difficult; Levine speculates that fast environments attract Type A people, while slower, Type B people tend to flee fast-paced cities.

The researchers went on to compare these results with the statistics for heart disease in the 36 cities and found a correlation between pace and disease even stronger than that between Type A behavior and heart disease in individuals. (The latter correlation is being refined. Many researchers now believe that it is not Type A behavior per se that leads to heart attacks but the degree of hostility involved.)

Levine then goes a further step. Cigarette smoking, a major risk factor for heart disease, shows the same regional pattern as do heart disease and a fast pace. Smoking, Levine says, is often related to psychological stress. One possibility is that stressful, time-pressured environments lead to unhealthy behaviors such as cigarette smoking and poor eating habits, he continues.

This is the straight stuff, complete with sentences such as "Our model of the fast-paced 'type-A city' may provide a basis for examining this hypothesis." But there is more to life than models, and as I read Levine I kept turning to a small book (written short, the author notes, to save the reader time) by Jean-Louis Servan-Schreiber, a French journalist, called *The Art of Time.* He talks about life at a fast pace, too, but with a different perspective: "For a long time I admired men in a hurry. Until I realized that they were merely under stress. *What I fear most about stress is not that it kills, but that it prevents one from savoring life.*"

Servan-Schreiber divides time in ways that textbooks don't. He begins with nature, "cosmic time" counted in years since the Big Bang or, on a more modest scale, the days and years we experience on our planet. "This time doesn't give a damn whether the planet has life on it," he says. Then there is society's time, imposed on us from the outside, "made up of interlocking codes and practices . . . omnipresent in our existence." Unless we understand the rules and management of social time, Servan-Schreiber says, we may be overwhelmed by it.

Finally, there is experienced time, better measured in terms of quality rather than quantity. Our perception of how fast time passes depends on what we are doing: It is one thing on our wedding night, another if our fingers are caught in the car door. Experienced time occurs when we are able to stop worrying about the past and the future long enough to live for a moment in the present, to feel and think and savor. It is this kind of time we feel lacking in our lives, Servan-Schreiber writes: "To some it is merely a memory."

It is experienced time that produces the sensation that time moves faster as we grow older. Servan-Schreiber compares life to a vacation. For the first couple of days at the cottage, the two weeks stretch away before us and we feel we have time to do all the things we want to. By the middle of the second week the end of the vacation is well above the horizon, and we know that there is not enough time to do it all, no matter how much we pack into the last days. Just so, of course, there comes a point in a life when death rises above the time horizon, and we realize that we do not have an infinite number of days to spend as we wish. Supposedly it is this realization that leads to the infamous "midlife crisis." I've been going through mine for 15 years now.

Servan-Schreiber describes how he learned the hard way to give time to time, to plan for a day or a decade, to treat the appointments he makes with himself as seriously as those he makes with other people. But it is not so much a how-to as an inspirational book, promising that there is a way to trade stress for serenity, even in the busiest lives. He struck too close to home when he asked: "How then does it happen that . . . Jill arrives on time, completes her work on schedule and appears relaxed while Jack runs after the minutes, misses his plane, and always finds reasons for procrastinating?" (When he talked about a disordered desk sucking all the energy out of a person, I wondered if he had been lurking about up here on the third floor of the Arts and Industries Building.)

He promises we all can learn to handle time like a martial arts master turning aside an attack with elegant economy of motion, like a virtuoso who no longer has to think about the bowing as he plays the violin. He offers the dream of a life so well organized that when old friends unexpectedly call from the airport to say they are in town overnight, it is possible with no stress at all to drop everything, join them for dinner, and get them back to the airport next morning.

The book is laced with aphorisms, many of which I found myself quoting over the telephone to family and friends. He frequently compares time to money, as in, "We think much more about the use of our money, which is renewable, than we do about the use of our time, which is irreplaceable." Or, "It is wiser to hand over our checkbook to someone else than to hand over our datebook."

Pointing out that everyone in the world has exactly the same amount of time, he asserts: "The paradox of time is that people rarely consider they have enough, when in fact all of it is available to everyone." My favorite comes in a section in which he is describing how many of us have settled into an unthinking routine: "Most of our work we do on automatic pilot. This represents progress in an airplane but not in us." His clinching proof that any of us can "make" more time when we want to is prototypically Gallic: "I venture to say you have probably been in love. A little, a lot, in any case enough so that, suddenly, in a life that was already full, you found yourself making long telephone calls several times a day, dreaming of the image of the other person, walking together aimlessly through the streets, leaving work early. So where did you find the time to do all that?"

After all this reading, and a modicum of reflection, I suddenly see how easy it is to seize more of those half-million minutes for myself, how to get more out of life, how to do all the things that until now I've only dreamed of. The secret, it turns out, is . . . . Oops. The production manager has appeared in the doorway. I'm out of time. Terribly sorry about that.

January 1991

Nibbling and Napping for Longevity

 Faint streaks of hope are lightening the long, dark night of denial and deprivation. After all these years of telling us that anything that tastes or feels good will kill us, those unsmiling people in the white coats have some good news. No longer will we be exhorted to live like Trappist monks, eschewing the pleasures of the table, treating sloth like the capital sin it once was. It now seems that some of our instincts are right on the money, that at least some of the time doing what feels good is the path to a longer, healthier life.

Regrettably, it is not yet the millennium. Woody Allen's fantasy of science discovering that steaks and cigarettes are good for us has not come true. The South Jersey breakfast special—creamed chipped beef, scrambled eggs, home fries from the mountain on the grill—is still off-limits. Hot dogs at the ball game remain a no-no. An egg salad sandwich on wheat toast with a chocolate milk shake, once the very model of a moderate lunch, must remain a dream right up there with lions on the beach.

Yet there is good news. Take the matter of between-meal snacking. It has been frowned upon for as long as I can remember. Only those of us weak of will stooped to such practice. Now we find out that eating 17 times a day is much better for us than the traditional three. A team of nutritionists, clinical biochemists, and research physicians at the University of Toronto came up with the good news. Nibbling all day lowers a person's cholesterol count by lowering the level of low-density lipoproteins (the LDL, or "bad cholesterol," component). The researchers have even figured out how this happens. The nibbling diet results in lower insulin levels. Insulin stimulates the production of an enzyme the liver needs to produce cholesterol; less insulin means less enzyme, which means less cholesterol.

Lower insulin levels also appear to directly reduce the risk of heart attack. Insulin stimulates the formation of fat in arterial tissue, and enhances the growth and proliferation of arterial smooth-muscle cells. The new cells narrow arteries even as fat deposits threaten to block them. Or, in the dry prose of medical research, "High . . . insulin levels are associated with subsequent [heart attacks] in nondiabetic men."

The benefits of nibbling are not a wholly new idea. Papers on the subject appeared as long ago as 1963; in 1970 a contribution to the *American Journal of Clinical Nutrition* was titled: "Meal frequency—a possible factor in human pathology." No one followed up on it, however, because the concern at the time was with overweight patients, and spreading three meals out into 17 did nothing for weight loss.

The Toronto researchers outnumbered their experimental subjects, 12 to 7. The seven healthy men were fed a normal, three-meal-a-day diet for a 2-week period, while maintaining their normal level of activity. During another 2-week period, exactly the same diet was fed to them in 17 separate snacks, one when they first woke up and then one every hour through the rest of the day. The results, published in the *New England Journal of Medicine*, were clear: While the seven were on the nibbling diet, their LDL dropped an average of 13.5 percent and their insulin levels 28 percent.

This bit of good news does involve some temptation of its own. An enthusiastic reader of the report might be tempted to combine the regimes, snacking 17 times a day in addition to the traditional three meals. (When Metrecal, a meal substitute, first appeared on the market, I remember reading about dieters who faithfully but mistakenly drank it before every meal, then complained that they were not losing weight.) But surely we can work out the details. And if I answer the telephone with my mouth full, it just means I am taking care of my heart.

The second bit of good news has to do entirely with yielding to, rather than resisting, temptation. Sleep researchers have belatedly discovered that human beings are programmed for an afternoon nap. For all the people who don't get quite enough sleep at night, a nap enables them to better concentrate on a task at hand and to make complicated decisions. For those who have gotten enough sleep, the main benefit of a nap appears to be an improvement in mood. Both would seem to be to the advantage of hard-nosed employers as well as to the individuals themselves.

Midafternoon drowsiness is not simply a function of a heavy lunch, it turns out. Roger Broughton, a neurologist at the University of Ottawa, says alertness and intellectual ability drop off in midafternoon whether or not a person eats lunch. For years, sleep researchers tried to keep their subjects from taking naps so they could better study what happens during sleep at night. (Had none of them ever felt a 4 P.M. slump? Ever had to fight off afternoon drowsiness so as to go on studying sleep?) But according to a report in the *New York Times* last year, they now feel they can no longer ignore the evidence from subjects who were wired or kept diaries or were kept underground for weeks at a time to see what patterns they established.

Those last provided the strongest evidence, not published until 1986. Scott Campbell and colleagues at the Max Planck Institute in Munich kept volunteers underground for weeks at a time with no clocks and no clues as to when it was really day or night. They found that the volunteers tended to fall into a 25-hour cycle, during which they slept once for the equivalent of a night's sleep and again for 1 or 2 hours. Even more interesting, they found that on average the naps began about 12 hours after the middle of the main sleep period. For a person who slept, for example, from 11 P.M. to 7 A.M., the midpoint was 3 A.M. The nap tended to begin at 3 P.M., 12 hours later.

Peretz Lavie at the Technion in Haifa has kept volunteers on a 20-minute sleep-wake cycle for several days at a time—they sleep for 7 minutes and stay awake for 13—to see how quickly, if at all, they can fall asleep at different times of the day. He has found that there is a peak in people's readiness to go to sleep in the midafternoon as well as at night. Before the Industrial Revolution made us all clock punchers, I would guess, the afternoon nap would have been part of the daily routine. It hangs on in some tropical countries, where the siesta is still tradition. (In Rome, that paragon of civilization, there are still four rush hours a day, two of them caused by workers going home for the 1-to-4 lunch break.)

So far, so good. When one of those pesky younger staff members comes bouncing into my office, all bright-eyed energy, and finds me (1) munching out or (2) meditating with my eyes closed, he or she will know that I am (1) being good to my heart or (2) making myself even more mentally alert and ready for those tough decisions.

And still there's more. An article in *Today's Chemist* goes into the biochemistry of decreasing one's percentage of body fat, and concludes that slow and easy works better than running flat out. During the first half-hour of exercise, the body tends to burn more carbohydrates (specifically, glycogen stored in muscles). But after about 30 minutes the body will begin to conserve glycogen and burn more fat, instead. The body "learns" to burn fat. The author, a retired chemist and long-distance runner named Trevor Smith, writes that three 1-hour workouts at low intensity will burn more fat than six 30-minute sessions at higher intensity.

I'm definitely the low-intensity workout type myself. Sometimes the intensity is so low that only a trained observer can tell I am working out at all. That faint aroma you can't quite place is the happy scent of burning fat. There is a long way to go, however; we need more good news. Our scientists are not doing enough. So here's my plan. First, I'll pull together a few tens of millions of dollars (this part is still a little vague). Set up a foundation to endow a brand-new research institute. Hire the best minds and set them to the task: Discover that the staples of editor-writer lunches, the foods that produced the ideas that made America's magazines great—eggs Benedict and bacon cheeseburgers—are so good for you that all thinking people will have them three times a week. Then, with that behind us, we'll go to work on ice cream.

May 1990

A Catheter in the Heart

Back in my glory days as a science writer, I wrote the occasional medical breakthrough story along with tales of subatomic particles, astronomical discoveries, and all the rest of "hard" science. To a young reporter, they were all the same: news stories with the added challenge of explaining just what it was you were talking about. Describing how surgeons slithered a balloon pump for a failing heart up an artery and into position was no different from explaining what happened in the racetrack we then called an atom smasher.

Today I know that there ought to be a law: Never write about something until it happens to you. My first clue came a decade ago. During the energy crisis, we ran a story about heating with firewood. At the time I thought it was a great idea: A public service to encourage people to heat with a renewable resource. Then the lawyers who were holding the 65 acres behind my Maryland house had the land timbered. That wasn't so bad. The holes in the canopy were probably a plus for the local wildlife, and I thought the "road" would heal. Next the lawyers invited the neighbors in to cut firewood. In no time a stretch of creek where great blue herons had once stalked by day and great horned owls had hunted by night looked like the site of a major tank battle. For a long time I could not bring myself to walk back there.

Twenty years after blithely writing stories about running catheters up people's arteries to their hearts, I have watched one invade my own. I have lain helpless on a table, watching on a screen as a tube thin as a wire works its way up my body to my heart, squirting dye like some irritable microcephalopod as it advances, finally spewing its inkiness into the left ventricle of my heart. The x-ray camera moves up and down my body, dancing from side to side in semicircular arcs like an automated lifeform analyzer aboard an alien spaceship. When the cath lab crew shoves on my groin to move the catheter, I go rigid with pain, at least until the fourth or fifth syringe of anesthetic. The dye feels like hot saltwater washing through my body. Coronary arteries look like downed power lines thrashing about in a storm. Until the dye reaches it, the heart itself is invisible; as it expands and contracts the arteries dance like Maypole ropes moved by invisible dancers. It is nothing whatsoever like being at a press conference or sitting at a typewriter back at the office. Not only is the whole procedure uncomfortable, the lab director has carefully explained to me that one possible side effect from this routine angiogram is death.

Strange but friendly faces from many parts of the world lean over me, telling me that I am lucky, that in up to half of all cases of heart disease the first symptom is death. They have a lot to say about my smoking, about how nicotine causes the walls of coronary arteries to spasm, how it makes blood "stickier" and thus causes it to clot, how it depresses the HDL or "good" cholesterol. I ask if they resent spending their days trying to save victims of self-inflicted disease—people like me who have smoked and eaten and lazed their way into the coronary care unit. Not the first time around, they tell me. It is only when a discharged patient goes back to the same old habits and ends up right back in the hospital that they begin to lose interest.

It is hard to explain to the pure of lung why anyone smokes in this day and age. The smoker knows better than anyone about lung cancer, emphysema, heart disease, gangrene in the extremities, sexual impotence—the whole dreary litany. The smoker knows all about social opprobrium, burned clothes, the fear that his house is on fire right now because he left a cigarette going.

Two months before my heart attack, it became harder for me to explain. A routine visit to the doctor turned up what look like spiderwebs in my lungs, a filigree of scar tissue that leaves the lungs too stiff for the diaphragm to assume its natural shape. (For a long time I assumed that the X ray was not mine at all but only one the doctor had bought from a supply house to scare smokers; now I'm not so sure.)

The condition is what doctors fetchingly call chronic obstructive pulmonary disease or COPD, a permanent loss of lung function.

A return visit for a stress test went no better. After a few minutes I had to jump off the treadmill, not for any pain in my chest or even because I was winded, but because a pain in my leg— a pain I have felt for years whenever I walk uphill—was now excruciating. I had assumed it was the pain of a middle-aged man who is totally out of shape. Not so, replied the doctor. It comes from smoking. The arteries in my legs are constricted, making me limp. Claudication, they call it.

Why keep smoking? Why, for that matter, have I spent that $20,000 for those 525,000 cigarettes? The short answer is that nicotine is the best all-around psychoactive drug on the market today. It gets you going, keeps you going, then relaxes you when you are ready. Smokers know that, but does the rest of the world really understand?

To the rescue comes Neal L. Benowitz, a medical doctor in the Clinical Pharmacology Unit at San Francisco General Hospital Medical Center, writing in the *New England Journal of Medicine*. He describes how fast nicotine gets to the brain, lists the "feel-good" neurotransmitters and hormones it releases. Benowitz says that in abstaining smokers, nicotine can improve "attention, learning, reaction time and problem solving" —everything a person could want. It is hard to know whether the improvement in performance and mood comes from the relief of withdrawal symptoms or from an "intrinsic enhancement effect on the brain," he adds, but at least some studies show the performance of nonsmokers improving after taking nicotine tablets, evidence for direct enhancement.

There's more: "Nicotine has been shown to increase vigilance in the performance of repetitive tasks and to enhance selective attention. Smokers commonly report pleasure and reduced anger, tension, depression, and stress." Not bad for a drug that is legal, cheap, widely available, easy to take and—at least for the longtime addict—tastes good. Imagine the director of research for some huge corporation who reports a find like that to his CEO.

Of course, there are those pesky drawbacks. It was one morning last November, the day before the Benowitz paper was published, that I was sitting at my desk, drinking coffee, smoking cigarettes, polishing sentences in a story for the next issue. A feeling of disengagement came over me, as though parts of my body were separating from one another, or my body and mind were going in different directions. Then a pain developed high in my chest, as though someone were holding a sharp pencil to my heart and pushing harder and harder. Pain spread across my chest; my left arm went numb. I tried doubling over; I tried leaning back; I tried imagining myself in a boat on Long Island Sound on a summer morning; the pain kept coming. As would 1.5 million other Americans last year, I was having a heart attack.

Later a cardiologist runs the angiogram film for me. The picture is black and white, jerky; the leaping arteries look like something you might see through a microscope. He stops the film to show me the circumflex artery, one of the secondary vessels that supply blood to the heart muscle itself. It looks to my eye as though a square notch has been cut out of it. I am looking at stenosis, a blockage, a place where plaque has built up on the side of the artery, cutting off 50 to 70 percent of the blood supply to one portion of my heart. It is this section that closed completely, either because it spasmed or because the plaque cracked and a clot formed. It is this one little spot in all the miles of blood vessels in my body that has me eating twigs instead of bacon and eggs, and keeping burglar's hours, out walking before sunrise and also when honest people have gone to bed. It is that one little spot, and the pencil-point pain it produced when I tried a cigarette after the hospital, that has me off the evil weed at last.

Three months later the excitement, the special attention, is over. This old body will never be the same, though. After 90 days without a cigarette, red blood cells are carrying oxygen again instead of carbon monoxide. Cilia, tiny hairs that line healthy throats, are breaking through the scar tissue in mine like weed trees coming up through the broken asphalt of an abandoned parking lot, ready once again to sweep away dust, dirt, soot from diesel engines, droplets of unburned jet fuel falling from the sky. Blood vessels that had been tightened down for years are relaxing, widening; climbing hills does not hurt anymore. The heart is working less and enjoying life more.

Some changes are unsettling. When you have not regularly breathed through your nose since the Dodgers played in Brooklyn and then start again, you feel like a new hole has opened in your head. You can smell the world around you once again; some is pleasant, some is decidedly not—how do dogs stand it? You face the prospect of no longer being able to lie around at home half the winter with the flu.

What if my immune system becomes bored? Will it miss the challenge of arsenic, carbolic acid, pesticide residues heated to hundreds of degrees? What is a body to do without its customary 40 doses a day of acetone, ammonia, benzene, boron, cadmium, creosol, DDT, endrin, formaldehyde, hydrogen cyanide, radioactive lead, mercury, methane, methyl alcohol, nitric oxide and the other 3,000 or so compounds present in cigarette smoke?

All the news is good, and yet . . . I still feel as though someone dear has died, as though I am swimming underwater and am desperate to take a breath. The urgency is permanent, like the sensation of falling that an astronaut has hour after hour, day after day. Something is wrong, something is missing. Every day I prove, in reverse, all the effects Dr. Benowitz found: I cannot think, cannot remember, cannot concentrate. The only thing more difficult than writing without smoking, I now discover, is to write about smoking without smoking. I am tapping along at 500 calories per paragraph: crackers, candy, cake, bread and jelly, soup, leftovers, anything a grazer can find.

In a recurring fantasy, it is night. I am sitting on the curb outside an emergency room. An open bottle of nitroglycerin tablets is at my side, in case a coronary artery spasms. When things seem to be calm inside the swinging doors, when that whole team of highly trained medical professionals is just waiting for something to happen, I take out a cigarette, light it, draw in deeply, more deeply . . . . In the fantasy, it is like mainlining Valium.

The books say you can expect about 48 hours of physical withdrawal symptoms: headaches, irritability, that sort of thing. I can report that in a scientific sample of one, the irritability continues for at least 90 days. It would not be a good idea right now for me to be at a press conference at which a tobacco spokesman asserts that cigarettes are not addictive, that a smoker can quit any time he or she decides to. I just might drop my notebook, gulp down a calcium-channel blocker or two, and then dispute that position. As Dr. Benowitz might say, I would attempt to destabilize the gentleman's electroencephalographic state. The spokesman could then write his story about hospitals, and I could write mine about jail.

April 1989

EPILOGUE
Recording a Life in Columns

 And so a life is measured out, not in coffee spoons but in columns. These essays are reports of forays, on foot or in the mind, by one individual who is still finding out what the questions are. As arranged here, they reflect the evolution of my interests in an unexpected way. They follow the same progression from the inanimate, astronomy and physics, to the nonhuman animate, biology and natural history, all the way to concern for human beings. That's a little simplified, of course. A husband and father working as a courthouse reporter all those years ago could be said to have had some interest in people. But there is a narrative flow from backyard astronomer to weekend birder to at least a passive (organization-joining, check-writing) kind of activist trying to influence the human impact on the planet.

The running theme of these columns is the central lesson of ecology, as articulated by, among many others, biologist and presidential candidate Barry Commoner: Everything *is* connected to everything else. It is the beginning of wisdom, an environmental ethic.

Learning to see means learning to appreciate. Just as a child grows from exclusive concern with itself to identification with family and friends and eventually the whole human species, just so the human race can grow to appreciate the nonhuman world—not for its potential value to us, but for its own sake. The central tenet of deep ecology is that all things in the biosphere have equal value and an equal right to live. All species use each other, of course, for food among other things, but the trick is to distinguish between what use is vital and what is unnecessary. Put another way, the goal is to live as lightly on the Earth as possible, taking only what we need.

In *The Voice of the Earth* Theodore Roszak wrote: "The ecological ego matures toward a sense of ethical responsibility with the planet that is as vividly experienced as our ethical responsibility to other people. It seeks to weave that responsibility into the fabric of social relations and political decisions."

This weaving is already happening, whether it be new environmental laws in this country or a new ecological awareness at the World Bank. Whether it is too little, too late remains to be seen. At the same time, even the people most concerned with those other nations, as Henry Beston called our brethren species, are striving to step ever more lightly. Arne Naess, the Norwegian mountain climber and philosopher who first articulated much of deep ecology, once agonized over whether he should walk in the woods, knowing that in the process he would inadvertently trample seedlings. Birders know that too much attention will drive brooding parents from the nest; whale watchers now worry about disrupting migration patterns.

This heightened sensitivity rebounds into our own human world. A persistent criticism of most nature writing is that it is misanthropic, either ignoring human beings altogether or seeing them only as destroyers. Often enough, the charge is a true bill. I'm as guilty as anyone. In an early column I tried to picture Manhattan reverting to nature after all the people were gone. I still daydream about a Utopian world in which social and economic justice prevail throughout Central and South America, so anyone could jump in a car and drive from the United States to the fabulous national parks there without fear of violence—or having to come face-to-face with dehumanizing poverty. The same kinds of questions arise here at home, of course. Is it OK to worry about losing migratory birds when we are losing children?

There is a politician on the national scene with whose philosophy and tactics I could not disagree more. Yet the ending of a speech I heard him give made a lot of sense to me. He challenged the group he was addressing to get involved in the political process, on whatever side and for whichever causes they choose. Invest 5 hours a week and 5 percent of your income, he said. Don't just lament: Do something.

Don't misunderstand. I have yet to take up that challenge, just as I have yet to put into practice more than a few of the ideas I have expressed in these columns. But it sounds right. I'll never stop walking in the woods, but the best thing I can do for them may be to stay out long enough to do my little bit to solve the problems that overshadow them and us. And even here there is an overlap between what might be called the ecology of the whole world, human society as well as the nonhuman biosphere, and personal ecology. People who know say working for a cause larger than one's self is a prescription for a happiness otherwise hard to come by. None of us can save the world. But we can give it a nudge. The particular cause is not the important thing. The working is.

UNIVERSITY PRESS OF NEW ENGLAND
publishes books under its own imprint and is the publisher for
Brandeis University Press, Brown University Press, University of
Connecticut, Dartmouth College, Middlebury College Press,
University of New Hampshire, University of Rhode Island, Tufts
University, University of Vermont, and Wesleyan University Press.

LIBRARY OF CONGRESS CATALOGING-IN-PUBLICATION DATA
Wiley, John P.
    Natural high / John P. Wiley, Jr.
      p.   cm.
    "Essays . . . originally published as "Phenomena" columns in
Smithsonian"—T.p. verso.
    ISBN 0–87451–624–2 (pa)
    1. Natural history. 2. Nature study. 3. Ecology. I. Title
QH81.W556   1994
508—dc20   92–56912

Design and production: Christopher Harris / Summer Hill Books